ESD
FROM A TO Z

ELECTROSTATIC DISCHARGE CONTROL FOR ELECTRONICS

John M. Kolyer

and

Donald E. Watson

VNR VAN NOSTRAND REINHOLD
New York

Library of Congress Catalog Card Number 89-70503
ISBN 0-442-00347-1

Printed in the United States of America

Van Nostrand Reinhold
115 Fifth Avenue
New York, New York 10003

Van Nostrand Reinhold International Company Limited
11 New Fetter Lane
London EC4P 4EE, England

Van Nostrand Reinhold
480 La Trobe Street
Melbourne, Victoria 3000, Australia

Nelson Canada
1120 Birchmount Road
Scarborough, Ontario M1K 5G4, Canada

16 15 14 13 12 11 10 9 8 7 6 5 4 3 2 1

Library of Congress Cataloging-in-Publication Data

Kolyer, John M., 1933–
ESD from A to Z: electrostatic discharge control for electronics
John M. Kolyer, Donald E. Watson.
p. cm.
Includes bibliographical references.
ISBN 0-442-00347-1
1. Electronic apparatus and appliances—Protection. 2. Electric discharges. 3. Electrostatics. I. Watson, Donald E., 1930–
II. Title.
TK7870.K633 1990
621.381—dc20

89-70503
CIP

Contents

Introduction

This book does not pretend to be an encyclopedic text, which would have to be a compilation of contributions from experts in various facets of ESD control. It is closer to being a handbook, but we feel that the most accurate descriptive term is an *approach:* a coherent, logical, and cost-effective system. As the building blocks of this system, about fifty real and conceptual tools from A to Z appear in **bold type** throughout the book in order to emphasize their usefulness and to provide an organizing principle. Insofar as "A to Z" hints of thoroughness, the thoroughness is not in the length of the A-to-Z list but in the applicability of our approach to all ESD-control situations.

This edition of *ESD from A to Z* differs from the self-published edition of 1989 in that various improvements have been made in the text; some of the papers in the Appendix have been condensed to remove extraneous data, and an extensive index has been added to make the book more useful.

The book is arranged as follows. First, Chapter 1 tells how to use the book. Then the real and conceptual tools are listed and explained in Chapter 2, and two of them are of such importance that they are given their own chapters: Chapter 3 for the static-safe package **(SSP)** and Chapter 4 for the static-safe workstation **(SSW)**. Since the best way to teach is by examples, the tools are put to use in Chapter 5 on ESD troubleshooting; this chapter is a collection of actual case histories illustrating the flexibility of our approach. This brings us to the bottom line: codification of our ESD control methods into specifications for packaging and for in-plant handling (Chapter 6) and into a program of maximum simplicity and minimum expense (Chapter 7). The important topic of disposition of mishandled hardware is discussed in Chapter 8. Finally, Chapter 9 wraps up the book with check lists of what to buy and do and a conclusion summarizing the elements of our approach, and Chapter 10 gives speculations on the future of ESD control.

References are listed at the end of the book, before the Appendix. As an example of the numbering system, Ref. 2–3 is the third reference cited in Chapter 2.

The Appendix contains our ten published papers or articles which give experimental justification for our rules and techniques as well as details of test methods and discussions of special topics such as tote boxes and the permanence of the **antistatic** property of **antistatic** materials. These publications are referred to throughout the text.

ESD control is notoriously controversial, and every subject, e.g., workbench surfaces, wrist-strap monitors, **ionization** systems, and protective bags, has its warring factions of opinion. We've never been shy on taking positions, and this book expounds our particular ways of doing things. Our approach is a combination of logic, worst-case but realistic testing, and simple, effective ESD-control procedures which are supported by extensive data in the Appendix and have proved themselves over many years of experience. Our object is to err on the "overkill" side of protection while still being cost-effective. We emphasize (and spend money on) the essentials and deemphasize (and save money on) the nonessentials. "Buy the basics, forget the frills." This attitude makes our approach especially suited to small companies with limited ESD-control budgets.

Throughout the text we use "**ESDS** (electrostatic discharge-sensitive) **item**" as a generic term including devices, subassemblies, and assemblies. Our program is designed for devices sensitive to 100 V by the Human Body Model (**HBM**). Separate programs for different sensitivity classes in one manufacturing area are impractical, in our opinion, so we recommend that the methods of this book be applied to the whole operation if any devices which are Class 1 by MIL-STD-1686A are being assembled, stored, or shipped.

In keeping with the practice of the EOS/ESD Symposia at which the papers in the Appendix were presented, trade names are given only for test equipment. ESD-protective materials and equipment may be found in buyers' guides such as those published by *EOS/ESD Technology Magazine* or *Evaluation Engineering Magazine*. The test equipment cited by trade name is endorsed for its function and design, but we imply no claims about durability, hazards, etc., nor do we suggest that patents be violated in buying or using the equipment. Giving brand names for other than test equipment would be futile because some products would be obsolete by the time the reader sees this book. In fact, the ever-changing nature of ESD-control products makes it necessary to append a "living list" of them, Reference Document 4, to Model Specification 1 in Chapter 6.

Please read and follow Chapter 1, "How To Use This Book," before delving into details. We wish you good luck with your program, but luck really shouldn't be necessary in what purports to be a systematic art if not quite a science. Armed with the arsenal of real and conceptual tools in Chapter 2, you'll solve ESD problems unique to your processes. Then, with well-trained personnel and a well-organized and well-managed program (Chapter 7), incidents of mishandled hardware will be rare. Under such conditions, ESD control can even be fun!

Chapter 1

How To Use This Book

First, take an hour to skim through the Appendix and read the abstracts and conclusions carefully. Note the locations of detailed information, e.g., description of test methods, for future reference.

Next, study Chapter 2. This gives the **Basic Rule** and the key concepts, including damage mechanisms, from which all else follows. When you come to the static-safe package (**SSP**) and static-safe workstation (**SSW**), stop and read Chapters 3 and 4. Remember that everything we're doing is aimed at detecting, avoiding, or preventing the two hazards of **fields** and **discharges.**

Study Chapter 5 to see how unique problem situations can always be handled by using the gamut of ideas and methods from Chapter 2. As problems arise, consult this chapter again to refresh your memory.

Use the model specifications in Chapter 6 as the basis for your own, adding more detail as necessary.

Use Chapter 7 as a guide when you're setting up a new program or improving an existing program.

Note the approaches in Chapter 8 for the disposition of mishandled hardware; you'll face this problem sooner or later.

Use the check lists in Chapter 9 when you're buying equipment, solving problems, initiating a program, etc.

Read Chapter 10 on the future for general information.

In summary, learn the location of information in the book so that it can be an ever-useful guide for troubleshooting, material/equipment buying, specification-writing, and program management.

When the book has been thoroughly understood, the reason for each provision of the model specifications in Chapter 6 should be evident. The reader should see how every rule, though based on limited data and more or less empirical, is intended to implement the **Basic Rule.** Remember that control measures may be primary (essential) or secondary (redundant or backup); for example, in Model Specification 1, requirement paragraph 3.6.11 for the **CD Rule** is a primary precaution, but 3.6.12.2 for unnecessary touching of leads and 3.6.12.3 for unnecessary triboelectric charging by body movement are secondary precautions. However, all regulations, whether primary or secondary, are meant to reduce the chance of a damaging **discharge** and to minimize **fields** and keep **ESDS items** out of intense **fields.** We emphasize again that as a whole system of ethics flows from the golden rule, our whole system of ESD control flows from the **Basic Rule.** All personnel should be taught this philosophy because people are inclined to disregard rules they don't understand. A successful program requires an enlightened work force. We recommend that your personnel either study this book or be trained with course material based upon it.

We emphasize that this book cannot be a simple "instruction manual" because ESD-control programs must be tailor-made to fit circumstances; see the Conclusion section at the end of Chapter 9. Therefore, you must absorb all the information and advice given herein and recast it into your own personal approach which will agree with ours in outline but differ in details. The bottom line will be your own specifications derived from the models in Chapter 6.

Chapter 2

Real and Conceptual Tools from A to Z

First, about fifty principles, rules, definitions, test methods, etc., are surveyed "from A to Z" under the heading "Real and Conceptual Tools: Definitions." This is not a glossary of ESD terms but a limited list of those that are especially important to our approach; definitions of other terms, such as abbreviations for types of devices, can be found in textbooks on electronics.

Second, the terms are elucidated under the heading "Real and Conceptual Tools: Discussion," and two of the most important ones, the **SSP** and **SSW**, are given chapters of their own, Chapters 3 and 4. These tools serve to troubleshoot in novel situations (Chapter 5), to develop specifications (Chapter 6), and to build a logical and thorough ESD-control program (Chapter 7). As mentioned in the Introduction, these key terms appear in **bold type** throughout the book.

ESD control has its complexities but is basically a simple matter of avoiding **discharges** and dangerously intense **E fields**—not all **E fields,** which would be impractical. (*Note:* In our usages, "**E field**" and "**field**" are synonymous.) If all objects in the **SSZ** are groundable and grounded, and if the **CD Rule** is followed, most of the ESD problems will have been solved. But the reader shouldn't be lulled into false confidence by this statement, because achieving these basic requirements of **grounding** and following the **CD Rule** is more easily said than done, human nature being what it is, and a residue of problems, exemplified by those in Chapter 5, will remain. These may be solved by diligent use of our conceptual tool kit.

Table 2-1 categorizes the tools as primarily applicable to either **fields** or **discharges** except when no distinction can be made. There is often much overlap; **humidification,** for example, reduces not only **fields** but also **discharges** (which may be visible sparks when the air is dry enough), but it is most often thought of as lessening the **static charge** on **nonconductors,** especially common plastics, to reduce the danger of the **FIM.** The point of the table is that every concept, material, or object in this chapter plays a

Table 2-1. Categorization of Tools.

APPLICABILITY	TOOL: REAL (AN OBJECT OR MATERIAL)	TOOL: CONCEPTUAL
Primarily pertinent to **fields**	**Antistatic** (materials)	**Apparent charge**
	Faraday cup	**CD Rule**
	Field meter	**E field**
	Insulator	**FFB**
	Nonconductive (materials)	**Field**
	Nonconductor	**FIM**
	Static-dissipative (materials)	**Humidification**
	Topical antistat	**Ionization**
		Shielding
		Static charge
		Static field
		Surface resistivity
		Triboelectric charging
		Voltage suppression
Primarily pertinent to **discharges**	**Antistatic lotion**	**ARTG**
	Conductive (materials)	**Capacitance**
	Conductor	**CDM**
	Continuous wrist-strap monitor	**DI**
	Personnel Voltage Tester	**Discharge**
	Static-limiting floor finish	**Discharge test**
	Zapflash	**Grounding**
		HBM
		Shunting
		Walk test
Equally pertinent to both **fields and discharges**	**Coupon**	**Analysis**
	Faraday cage	**Basic Rule**
	MOSFET	**ESDS**
	SSP	**ESDS item**
	SSW	**Latent failure**
	SSZ	**Operator disciplines**
		Special test
		Standard test

role in preventing **DI** or the "internal **DI**" caused by induction ("pushing" or "pulling" of electrons by a **field** to create polarization within a device).

Real and Conceptual Tools: Definitions

Analysis: A paper study with no testing required.

Antistatic: Having a **surface resistivity** of at least 10^9 but less than 10^{12} ohms/square (our definition; same upper limit as EIA 541 "dissipative").

Antistatic Lotion: Hand lotion applied to the wrist to lower skin resistance and help meet **ARTG** for operators wearing wrist straps.

Apparent Charge: The charge in volts of a surface or object as read by a hand-held **field meter.** The word "charge" is used for convenience; the correct units of charge are coulombs (or usually nanocoulombs for ESD work).

ARTG: Allowable resistance to ground for operators' skin. For **ESDS items** sensitive to 100 V this is 10 megohms.

Basic Rule: An **ESDS item** shall never, even for a nanosecond, be exposed to an **E field** of more than a prescribed strength or have its sensitive leads or terminals touched to, or receive a **discharge** from, any surface at more than a prescribed voltage.

Capacitance: The capacity or ability of an object to hold a charge. In ESD control, the usual unit of **capacitance** is the picofarad, pF.

CDM: Charged device model, a damage mechanism.

CD Rule: Charge-distance rule. An empirical relation of safe distance from a surface versus the **apparent charge** on that surface. This rule protects items sensitive to 100 V from damage by **E fields** by the **FIM.**

Conductive: Having a **surface resistivity** of less than 10^5 ohms/square.

Conductor: A **conductive** material. Common examples in ESD control are metals and plastics loaded with graphitic ("metallic") carbon.

Continuous Wrist-Strap Monitor: Equipment for continuously checking the resistance of an operator to ground to make certain that the **ARTG** is never exceeded even for a nanosecond as required by the **Basic Rule.** A **continuous wrist-strap monitor** system includes the wrist strap with dual-conductor cord.

Coupon: An exaggeratedly ESD-sensitive model for an **ESDS item.** Coupons are passed through an automated process to locate hazards and, when these have been corrected, to give assurance that the process remains ESD-safe.

DI: Direct injection, a damage mechanism. (*Note:* This and the **CDM, FFB,** and **FIM** are called "damage mechanisms" as opposed to "failure mechanisms," which are the ways an ESD-affected device fails internally, e.g., dielectric breakdown [DoD-HDBK-263, p. 20]).

Discharge: The flow of electrons between surfaces. This flow may take place between touching surfaces or across a gap via ionized air as a visible spark or an unseen silent discharge.

Discharge Test: A test in which a finger of a charged person, a tool held by a charged person, or a probe connected to a charged capacitor is discharged to the surface of an ESD-protective container, e.g., a bag or tote box, holding a **MOSFET** in a special fixture. Damage to the **MOSFET** constitutes failure of the test.

E Field: Electrical field, which is the region surrounding an electrically charged object in which another electrical charge can be induced and will exert a force.

ESDS: ESD (electrostatic discharge)-sensitive.

ESDS Item: ESD-sensitive device, component, subassembly, or assembly. Used as a general term for convenience.

Faraday Cage: A container whose wall is electrically continuous and has at least the electrical conductivity of 0.00025-inch (0.25-mil) aluminum foil. Small holes, such as made by staples, are allowed. Wire screen, e.g., FED-SPEC-RR-W-365, Type VII, 18 × 16 regular, is an excellent **Faraday cage,** equivalent to at least 6-mil foil. (*Note:* This is our own definition.)

Faraday Cup: A shielded cup with attached electrometer for measuring the charge on any object dropped into the cup.

FFB: Field-from-board damage mechanism.

Field: An **E field** in the context of this book.

Field Meter: An instrument for measuring **E fields.**

FIM: Field-induced model, a damage mechanism.

Grounding: Metallic connection with the earth to establish zero potential with respect to earth. Grounds for ESD purposes, e.g., for an **SSW,** include water pipes and building structural steel. **Grounding** also means the electrical connection of **conductive, static-dissipative,** or **antistatic** materials with earth in any manner.

HBM: Human body model, a damage mechanism (special case of **DI**).

Humidification: Controlling relative humidity, e.g., to 30% or 40% minimum at 70°F, to reduce the **triboelectric charging** of materials in the **SSW.**

Insulator: A **nonconductor.**

Ionization: The ionization of air for the purpose of neutralizing charged surfaces.

Latent Failure: Failure of an **ESDS** device because of previous degradation and weakening by one or more ESD events.

MOSFET: Metal-oxide semiconductor field-effect transistor. Used in the **discharge test** and in **special tests.**

Nonconductive: Having a **surface resistivity** of at least 10^{12} ohms/square (our definition, consistent with that of the Electronics Industries Association).

Nonconductor: A **nonconductive** material.

Operator Discipline: A rule of conduct for operators that tends to minimize the chance of ESD damage to **ESDS items** being handled.

Personnel Voltage Tester: Trade name for a voltmeter with high input impedance and low output **capacitance,** for measuring the instantaneous voltage on people.

Shielding: Attenuation of an **E field** by an object between the charged surface and a point in space at which the **field** strength is measured. The more **conductive** the object blocking the **field,** the better the **shielding.**

Shunting: "Shorting out" or connecting leads together as a means of lowering the ESD sensitivity of an **ESDS item.**

Special Test: A test designed to resolve a specific ESD-problem situation. Typically, the test conditions are worst-case, yet realistic, so that confidence is high if the test is passed.

SSP: Static-safe package. This is an **SSZ** within a **Faraday cage** or within material providing sufficient **shielding** to pass a **discharge test** on the particular configuration of shipped/stored **ESDS item** and packaging.

SSW: Static-safe workstation. This is a volume in space which is designated and equipped to maintain an **SSZ.**

SSZ: Static-safe zone. A volume in space at every point of which the **Basic Rule** is followed.

Standard Test: A test procedure from a specification or other ESD-control document.

Static Charge: The quantity of electricity, measured in coulombs, nanocoulombs, etc., accumulated on a **nonconductor.** This is a classic definition, but **static charges** occur, of course, on electrically isolated (ungrounded) **conductors** and exist for a brief time, but drain off quickly, on grounded **antistatic** or **static-dissipative** materials.

Static-Dissipative: Having a **surface resistivity** of at least 10^5 but less than 10^9 ohms/square.

Static Field: An **E field** caused by a **static charge.**

Static-Limiting Floor Finish: Floor "wax" with a **static-dissipative** or **antistatic** surface as well as a propensity to reduce **triboelectric charging** of people walking or shuffling their feet while sitting.

Surface Resistivity: The electrical resistance between two electrodes pressed against a surface and forming the opposite sides of a square of any size. The units are ohms/square, and the basic reference is ASTM D257.

Topical Antistat: A moisture-attracting substance applied to the surface of **nonconductors (insulators)** to make them **antistatic.**

Triboelectric Charging: Generation of **static charges** by friction or separation of dissimilar materials or even two pieces of the same material.

Voltage Suppression: The "collapse" or reduction in strength of an **E field** around a charged surface when that surface in close to another surface, especially a grounded **conductive** one.

Walk Test: Realistic test for approximate voltage attained on walking persons by **triboelectric charging** between shoe soles and floor surface.

Zapflash: Trade name for a continuity checker with a bulb which glows dimly at about 7 megohms and becomes brighter with decreasing resistance.

Real and Conceptual Tools: Discussion

Analysis: Conclusions are arrived at logically, using methods such as extrapolation and criteria such as similarity. For example, we know that 0.25 mil of aluminum foil is a sufficient **Faraday cage** for general use, so extrapolation tells us that a ⅛-inch wall of aluminum on an electronic box is even better and need not be tested for **shielding.** Or, if one **antistatic** plastic composition with a **surface resistivity** of 10^{10} ohms/square bleeds off **static charges** satisfactorily, we can conclude that a different composition with similar **surface resistivity** may be substituted (assuming no shortcomings in other properties).

Antistatic: DoD-HDBK-263 defines "**antistatic**" as having a **surface resistivity** of less than 10^{14} ohms/square. However, our measurements with blade electrodes and a megohmmeter (Paper No. 2 in the Appendix), or with a Voyager SRM-110 **surface resistivity** meter, showed that values of 10^{12} ohms/square or more are too high for **static charges** to bleed off within a few seconds as desired. Exactly what bleed-off time is needed is a moot question, but we insist that **antistatic** surfaces not retain over 300 V **apparent charge** when stroked with polyester fabric or Aclar (polychlorotrifluoroethylene) film and held in one hand while being measured with a **field meter** in the other and waiting 5 seconds. Numbers aside, the **antistatic** surface must be sufficiently conductive for a feeble current to flow across it (Ref. 2–1).

(*Note:* The word "conductive" in the last sentence is not in **bold type** because it is not used in our special sense of having a **surface resistivity** of less than 10^5 ohms/square.)

Antistatic Lotion: Some operators have skin contact resistance above 10 megohms, which exceeds the **ARTG** (see Paper No. 10 in the Appendix), so this lotion is a necessity for them, and it may have to be applied twice a day. Thus the **continuous wrist-strap monitor** may become a lotion monitor, signaling when the lotion must be renewed. Fortunately, female operators enjoy using the lotion, which has a pleasant feel and can be obtained with either herbal or floral essence. Select a brand free of lanolin or silicones that might contaminate surfaces and interfere with adhesion for bonding or printing.

Apparent Charge: This is the uncorrected reading of a **field meter,** in volts, taken for a surface following the meter manufacturer's instructions. This reading reflects the intensity of the **E field** at the point in space where the sensor plate of the meter is held. The "voltage" of a **nonconductive** surface is of doubtful meaning because the **static charge** may vary greatly from spot to spot: excess electrons, or regions of electron deficiency, may exist in clumps. Besides, the surface charge per se is irrelevant. What matters is the strength of the **field**, but this is hard to measure even at one point because the meter itself perturbs the **field** and affects the reading. Also, the lines of force of the **field** are warped by nearby surfaces. However, despite these uncertainties the **apparent charge** is meaningful for our **CD Rule** because this rule is a worst-case approximation with latitude for numerical error. See **"Field Meter."**

ARTG: An ARTG of 10 megohms is recommended in Paper No. 10 in the Appendix. Note that, without the worst-case situation of a garment sleeve stroking a workbench surface, resulting in a very rapid voltage rise, the **ARTG** might be set much higher than 10 megohms. Operators' sleeves are supposed to be of **antistatic** smock material or else rolled up (see Example 3, "Operator Clothing," in Chapter 5), so this worst-case condition should not exist. However, we want backup precautions; if the law on rolling up sleeves is violated, an **ARTG** of 10 megohms still provides protection. Our program is designed to provide safety nets in this way whenever they are cost-effective. Another good example is **static-limiting floor finish.**

Basic Rule: For **ESDS items** sensitive to 100 V by the **HBM,** this rule may be rewritten: "An **ESDS item** shall never, even for a nanosecond, be exposed to an **E field** in violation of the **CD Rule** or have its sensitive leads or terminals touched to, or receive a **discharge** from, any surface at more than 50 V."

"Even for a nanosecond" refers to the rapidity at which ESD damage can occur. Only 25 V applied for 100 nanoseconds, at the chip, can destroy memories or microprocessors (Ref. 2–2). In the waveform required by MIL-STD-883, Test Method 3015, referenced in MIL-STD-1686A, the rise time for the **HBM** is a maximum of 10 nanoseconds (Ref. 2–3). A typical ESD event involves rapid **discharge** with rise times of 1000–3000 volts/nanosecond (Ref. 2–4).

Although a charge of 300 V on **nonconductors** is harmless for devices sensitive to 100 V by the **HBM** (Figure 3 of Paper No. 3 in the Appendix), it is usually not practical to distinguish **nonconductors** from other materials, and **nonconductors** shouldn't be in an **SSW**. Therefore, 50 V is stated as the limit for any surface.

A basic rule suggested by a supplier of **conductive** ESD-control materials is: **ESDS items** "may be handled or approached only with highly **conductive** grounded objects," and "storage or transportation can only be done in Faraday cages (maximum of 10^4 ohm/square material)." This is the "**conductive** approach." Note that our definition of **Faraday cage** is more rigorous; see the discussion.

In contrast to the **conductive** approach, a supplier of **antistatic** materials has long insisted that all surfaces in the **SSW** must be "nonsparking." This is the "**antistatic** approach," and its merit versus the **conductive** approach is a notable controversy in the very controversial field of ESD control. Which of these two approaches is right? The **conductive** approach is correct in theory but lacks the "forgivingness" or latitude for error needed in imperfect, real-life situations. Specifically, the **CDM** and **DI** are not taken into account. **ESDS items,** materials, or even careless operators might be charged, and **conductive** surfaces can increase ESD damage by rapid **discharges** as discussed under **"CDM"** below. Also, **conductive** items, e.g., bags, can carry lethal currents if a 110-volt hot lead falls onto them (Paper No. 7 in the Appendix); this danger is improbable, but safety is so important that even unlikely possibilities must be considered.

There are also other arguments against **conductive** materials, such as sloughing of metal flakes from metallized bags or particles from carbon-loaded plastic bags or tote boxes; these **conductive** flakes or particles could cause shorts by falling into open devices or onto printed circuit boards to bridge circuit lines. On the other hand, **antistatic** materials of the conventional type (depending on a fugitive antistat) have a permanence problem (Paper No. 2 in the Appendix), and their rubbed-off or volatilized antistats can cause contamination (Paper No. 9 in the Appendix). Nothing is perfect.

On balance, we use and recommend a mix of **antistatic, static-dissipative,** and **conductive** materials, as stated just before the "Conclusions" section of Paper No. 3 in the Appendix. We prefer to avoid **conductive** materials (note that **"conductive"** is defined by the conductivity of the *surface,* so an externally **antistatic** foil-laminate bag, for example, is not **conductive**), but we use them when no alternatives are conveniently available. After all, **conductive** items such as screwdrivers and other hand tools are ubiquitous in **SSWs**. We merely try to minimize the addition of more **conductors.**

In conclusion, the **Basic Rule** guards against **fields,** by means of the **CD Rule** and control of **triboelectric charging,** and also against **discharges,** by means of **grounding** of materials and operators, by **operator disciplines,** and by using **coupons** to make automated processes ESD-safe.

Capacitance: **Capacitance** C is related to the charge Q and the voltage V of an object by the equation: $Q = CV$. This means that for a given Q, or quantity of electrons present or missing, a decrease in C, which is the capacity to hold the charge, results in a rise in V, which is the "looseness" of electrons (or "anxiousness" of their lack) that manifests itself as an **E field.** Thus, when the capacity to hold the charge decreases, the charge is less tightly gripped or restrained and radiates as a **field;** when the capacity increases, this **field** collapses.

In practice, a charged object is one plate of a condenser (capacitor) separated by air as the dielectric layer from a ground plane such as a workbench surface or a floor, which is the other plate of the condenser. When the distance between the plates doubles, Q remains the same, C is halved, and

V doubles. This relationship is discussed in connection with the **ARTG** in Paper No. 10 in the Appendix. The most familiar, and important, example of V rising as C falls is a charged person lifting his foot.

CDM: For a discussion of CDM, see the complete version of Paper No. 8 condensed in the Appendix. As mentioned in that paper, the nature of the **CDM** has been questioned, with the suggestion that damage occurs in the charging process rather than during a **discharge,** so that the **CDM** resembles the **FIM.** However, our observations on the **FFB,** which is a kind of **CDM,** indicate that a **discharge** isn't necesary for ESD damage but increases its likelihood (Paper No. 8 in the Appendix).

Insofar as **DI** is a part of the **CDM,** a **conductive** surface tends to be more damaging than an **antistatic** surface (Papers No. 3 and 7 in the Appendix). This was demonstrated in tests which are admittedly contrived and probably seldom simulated in practice, but they support the intuition that an **antistatic** surface slows the rate of a **discharge** and is at least slightly safer than a **conductive** surface. Hence we discourage the use of **conductive** surfaces but don't forbid it; see **"Basic Rule"** above.

CD Rule: This empirical rule is based on **MOSFET** tests as described in Paper No. 3 in the Appendix. Since the **MOSFET** board used in the tests is an exaggeratedly sensitive design (large "antennas"), the **CD Rule** is worst-case, and this worst-case quality relieves uncertainty in **field-meter** readings as explained under **"Apparent Charge"** above. The equation $d = V^{0.5}/1.8$, where d is the minimum safe distance in inches and V is the **apparent charge** in volts, can be expressed as a table as included in Model Specification 1 in Chapter 6.

The **CD Rule** is incorporated in a company in-plant-handling specification, as it is in Model Specification 1, and several years of experience have proved this rule to be extremely useful. It has helped solve many problems, a few of which are described in Chapter 5. The rule is especially valuable in preventing needless expense; fears of low-intensity **fields** are discredited so that purchase of equipment to control these unreal hazards is avoided. For example, **conductive** chairs were shown to be unnecessary in Example 5 in Chapter 5. In other words, the rule forestalls "overkill" and maximizes cost-effectiveness.

Based on the monetary savings provided by the **CD Rule, field meters** should be generously distributed to operators. One **field meter** for each **SSW** is ideal. Not only do the meters allow cost-effective solutions to ESD problems via the **CD Rule,** but they are a prime tool of eternal vigilance to exclude common plastics and other sources of **fields** from the **SSW.**

In conclusion, the **CD Rule** will be one of your best friends in ESD control. Use it often.

Conductive, Conductor: Conductivity is defined by **surface resistivity**

because surfaces tend to be the locus of currents in ESD-control work; examples are vapor-deposited metal layers or antistat sweat layers. In the test (ASTM D257 or equivalent), current flows through the bulk of a **conductive** material such as carbon-loaded polyolefin, not just on the surface, but what matters from a practical standpoint is not where the current flows but how large it is, because the property of real interest is the rate of bleed-off of **static charges.**

Incidentally, this facility to carry current may be too large for safety; see **"Basic Rule"** above.

For this reason of safety and because of slightly more danger to **ESDS** devices (see **"CDM"** above), the use of **conductive** surfaces is discouraged but not prohibited in our approach.

Do not use a **field meter** to measure the true voltage on a **conductor.** The reading probably will be deceptively low, so that **DI** may be unrecognized as a threat. See the precautions under **"Field Meter."**

Continuous Wrist-Strap Monitor: This subject is discussed in detail in Papers No. 5 and 10 in the Appendix. The **Basic Rule** requires continuous monitoring "even for a nanosecond," and if a wrist-strap system is found to have failed in a daily check, for example, the operator's skin voltage may have been sufficient to cause damage by **DI** for not just a nanosecond but for a whole shift!

More and more **continuous wrist-strap monitors,** of the resistive type as we require, are being used in our assembly areas. There is no logical alternative, especially for high-reliability products in which **latent failures,** however few, could be disastrous. However, try monitors before buying to be sure the audible alarm is not annoying, etc.

Coupon: Paper No. 8 in the Appendix describes the **coupon** aproach and its successful application to an automated assembly process. The complete version of this paper also mentions damage mechanisms and ESD hazards.

DI: This is the most straightforward and common cause of ESD damage, with people being the major threat. Therefore, controlling voltage on personnel (Paper No. 10 in the Appendix) is the most important part of an ESD-control program.

The **HBM** is a model for **DI** caused by people, but the conditions are not realistically worst-case; see **"HBM"** below.

Discharge: **Discharges** and **fields** are the two ESD threats controlled by the **Basic Rule.** A **discharge** may be very rapid, e.g., with a rise time of 1000–3000 volts/nanosecond, or slow, e.g., 1000 volts/minute, in which case it is called "bleed-off" or "drain." **Conductive** surfaces favor rapid **discharge,** and there is evidence that these surfaces are therefore more hazardous to **ESDS items** than are **antistatic** surfaces (see **"CDM"** above).

A **discharge** is a flow of electrons, with the quantity of electrons being measured in coulombs and the rate of flow (current) in amperes = coulombs/second. A **discharge** creates a **field** which may itself cause ESD damage.

Discharge Test: In variations of this test, we use a resistance of 1500 ohms as in the **HBM** (Chapter 6, Model Specification 2, para. 4.3.3, option 1), 150 ohms (same reference, option 2), or 0 ohms (Test No. 1 in Paper No. 4 in the Appendix). For tote boxes, electrodes are used instead of a capacitance probe (see under "Test Methods" in Paper No. 7 in the Appendix). The test becomes much more rigorous as the resistance drops.

The **shielding/discharge** test with a capacitive sensor and oscilloscope (see Fig. 2 of Paper No. 3 in the Appendix) has become a standard for bag materials (in EIA 541, "EIA" being the Electronics Industries Association), but this test is misleading as explained in Paper No. 3 in the Appendix. The flaws in this test are (1) it gives falsely high (good) readings when metallization is exposed rather than buried, and (2) the flat electrodes don't provide a spark **discharge.** A "real-life" spark from a person's finger evaporates the metal from see-through metallized bags so that they afford poor protection; see Table 3 of Paper No. 4 in the Appendix. An analogy of an unrealistic test that is still on the books but has been largely supplanted by more realistic tests is ASTM D635 for flame resistance of plastics. Nylon passes this test, but glass-reinforced nylon, which if anything should be less flammable, fails. Why? Because the low melt viscosity of nylon allows the molten region to drip off, carrying the flame with it, unless glass fibers support the melt so that the flame can propagate! The shielding/discharge or pulse test of EIA 541 is just as unrealistic. One correction for ASTM D635 was to add cotton below the burning specimen with a provision that burning, falling drops must not ignite the cotton for the test to be passed. Similarly, a correction to the pulse test would be to add a spark **discharge** as we have done in our modified version. See Chapter 4 for further discussion of unrealistic versus realistic tests.

It is interesting that researchers at a major corporation producing see-through metallized bag material noted evaporation of the metallization by a spark and even studied the area evaporated versus voltage of the **discharge,** but this information was suppressed because it might have harmed sales. Our position is that spark **discharges** are a significant threat and bags should resist them.

E Field: **Fields** are, of course, the threat, along with **discharges,** that the **Basic Rule** controls. For practical purposes (the **CD Rule**), **field** strength is measured in terms of **apparent charge** and the distance from the charged surface. See the discussion under **"Apparent Charge"** above.

ESDS, ESDS Item: **ESDS** devices fail in two common ways. Voltage-induced failure dominates for discrete MOS (metal-oxide semiconductor) devices. A voltage of 60–70 V exceeds the dielectric-breakdown strength of the common thickness of 1000 angstroms of oxide. After breakdown, the part usually shows permanently degraded leakage characteristics and is more susceptible to future damage; it has a reduced operating margin and is likely to fail.

The other common failure mode is current-induced failure. This appears in Schottky or *PN* (bipolar) junction ICs, which are subject to destructive reverse breakdown in which the instantaneous power of the **discharge** (more than 5 kW for less than one microsecond) generates enough heat to melt silicon and metal. Thin-film resistors are damaged by surface breakdown that bypasses part of the resistor and leaves a shunt path, or else metal is vaporized or melted from the thin film. In either case the resistance value usually shifts, which is serious in a precision (less than 0.1% variation) part (Ref. 2–5).

Typical failure distributions for bipolar ICs are 90% junction burnout and 10% metallization burnout. For MOS ICs, this changes to 63% metallization burnout and 27% oxide punchthrough (Ref. 2–2). For a full discussion of failure mechanisms see DoD-HDBK-263, pp. 12–23.

MIL-STD-1686A, which has superseded DoD-STD-1686, defines three classes of ESD-sensitivity: Class 1 (0–1,999 V), Class 2 (2,000–3,999 V), and Class 3 (4,000–15,999 V). Class 1, which is the concern of this book, includes microwave devices (such as Schottky barrier diodes), **MOSFETs**, surface acoustic wave (SAW) devices, junction field effect transistors (JFETs), charged coupled devices (CCDs), precision voltage regulator diodes, operational amplifiers (OPAMPs), thin-film resistors, integrated circuits (ICs), hybrids using Class 1 parts, very high speed ICs (VHSIC), and some silicon controlled rectifiers (SCRs).

Note that components mounted on an assembly are relatively ESD-safe only if the assembly has protection circuits at all sensitive nodes. The risk of ESD damage can actually increase for mounted devices because each printed conductor connects to several devices and a **discharge** to that conductor stresses several devices, not just one. (*Note:* The word "conductor" in the last sentence is not in **bold type** because it is used in the special sense of "circuit line.") Another factor is that circuit lines can act as antennas to intensify the effect of **fields** and promote damage by the **FIM.** Furthermore, CMOS (complementary metal oxide semiconductor) circuits subjected to ESD while powered are subject to an additional risk called "latchup," which is an avalanche effect causing overheating and catastrophic failure (Ref. 2–5). We repeat: ESDS devices mounted on boards can be at even more risk than when unmounted (Ref. 2–6).

Faraday Cage: As defined in technical dictionaries, a "Faraday cage" is a room enclosed with an earth-grounded metal screen, and the only commercial ESD-protective bag in accordance with this defintion (except for **grounding**) is constructed of aluminum wire screen (as used for windows) sandwiched between layers of **antistatic** polyethylene (see Paper No. 4 in the Appendix). However, for most purposes 0.25 mil of aluminum foil, as in commercial foil laminate bags (see Table 10 of Paper No. 3 and Table 3 of Paper No. 4 in the Appendix) is sufficient. We recommend these foil laminate bags for all in-plant handling as well as shipping (but nonfoil package constructions may meet the **discharge** test of Model Specification 6). However, consider the screen bag when visibility of the contents is needed and extremely high ESD sensitivity, high cost, or need for high reliability of the enclosed item justifies a relatively expensive package.

Note that being a **Faraday cage** or equivalent is not enough for an **SSP,** because the inside of an **SSP** must be an **SSZ** in which **triboelectric charging** cannot violate the **Basic Rule** by creating hazardous **fields.** MIL-B-81705, Type I, is a perfectly good **Faraday cage,** but its **nonconductive** liner is unsuitable for an **SSZ;** in a test, miniature circuit boards with **MOSFETs** were shaken in a bag and found to be damaged due to **triboelectric charging** (Ref. 2–7). Our own tests verified that this liner was **nonconductive.**

In conclusion, use foil laminate bags with **antistatic** surfaces both inside and outside, or, in special cases, use screen bags with **antistatic** plastic sandwiching the screen (see Paper No. 4 in the Appendix).

Faraday Cup: For examples of the use of this test equipment in studying the propensity for **triboelectric charging,** see papers No. 6 and 7 in the Appendix. A **field meter** measures charges indirectly and imperfectly, whereas the **Faraday cup** gives accurate readings on any charged object dropped into it. However, when the intensity of the **E field** is the property of real interest, as it in using the **CD Rule,** the relevant parameter is the **apparent charge** read by a **field meter** and expressed in volts.

FFB: This damage mechanism is a special case of the **FIM. Triboelectric charging** of circuit boards, e.g., by sliding on stainless-steel conveyor belts or oven shelves, creates a **field** which damages **ESDS** devices by induction as explained under **"FIM"** below.

Field: See **"E Field."**

Field Meter: This has been and still is the single most useful tool in ESD control.

Field meters commonly are calibrated at the factory with a charged 1-ft^2 plate, but in actual measurements the target areas vary, with a large effect on the readings (Ref. 2–8). Thus the actual voltage, or charge, on the surface can't be accurately determined, but what matters for our purposes is the **apparent charge** as discussed under **"Apparent Charge"** above.

Though the real charge is irrelevant, we do want a properly determined

apparent charge as a measure of **field** strength, and there are a few requirements for using the meter correctly.

First, the meter must be calibrated. This is suitably done annually, with a calibration sticker being placed on the meter.

Second, if the meter has a grounding outlet it must be grounded with a lead, or if the meter has no such outlet the operator must be grounded.

Third, the meter must be zeroed on a grounded **conductive** object, suitably the grounded operator's hand. Note that meter readings are relative. If neither operator nor meter is grounded and the operator is charged to −1000 V and has zeroed the meter on his hand, he will read a grounded 1-ft^2 surface as +1000 V, or he will read a surface at +1000 V as +2000 V. Thus, to purposely read the approximate charge on himself, the ungrounded operator can take a reading on a grounded 1-ft^2 surface and reverse the polarity. Or a larger surface can be used for a rough result.

Fourth, the **apparent charge** must be read according to the meter manufacturer's instructions; for example, the meter might read true at 6 inches from a surface, or the distance might be increased to 12 inches, to keep the needle on scale, and the reading doubled.

We recommend that every **SSW** have a **field meter** to be used in complying with the invaluable **CD Rule.** However, the meter must be used only for **apparent charges.** It is *not* to be used for reading voltages on **conductors** because the true voltage is not shown by the meter unless the conductor happens to be a 1-ft^2 plate (assuming the meter was calibrated with this). For example, because of its small area a wire at 500 V might read only 50 V on the meter. According to the **CD Rule,** this reading of an **apparent charge** of 50 V prescribes a safe distance of 4 inches, which is fine, but a requirement of the **Basic Rule** for **ESDS items** sensitive to 100 V is that a sensitive lead may not touch a surface at more than 50 V in *real* voltage. In this example, if the wire were touched the item would be damaged by **DI.** In other words, **field meters,** as their name indicates, are for judging **fields,** via **apparent charge** and distance, *not* for measuring true voltages! To measure voltages on **conductors,** use a contact or noncontact voltmeter. A noncontact voltmeter is a type of **field meter,** but it is specially calibrated and applicable only to certain sample geometries; it is not an ordinary, general-purpose **field meter.**

Operators should be given detailed instructions on using the **field meter** because it is such an important tool. For familiarization, they should charge a sheet of **nonconductive** plastic, e.g., polyethylene, by rubbing with polyester or woolen fabric, and note the unevenness of the voltage readings over the surface because electrons or electron-deficient areas are clustered irregularly; in such cases, the highest reading, which is the most conservative for protection purposes, shall be used as the **apparent charge** in implementing the **CD Rule.** Also, operators should rub the plastic against grounded stainless steel and note that grounded **conductors,** though they never show a charge themselves because it immediately drains off, are very effective in charging **nonconductors** including circuit-board laminates. Intuition may suggest that grounded metals are harmless in regard to ESD, but this idea is dangerously wrong! Not only can they cause **triboelectric charging,** but they can be targets for rapid **discharges** in the **CDM:** see "**CDM**" above. The skin is another **conductor** that charges **nonconductors** by rubbing, as the operator can demonstrate by wiping a hand over the surface of an uncharged plastic sheet. (Use an ionizing blower to neutralize the plastic between **triboelectric charging** tests, or else the charge may be "blotted" off by pressing the palm of the hand, without rubbing, against the plastic surface.)

If the sensor of the meter is accidentally touched to a highly charged **conductive** surface, the meter may lock up and refuse to be zeroed; if this happens, give the meter time to recuperate. Sometimes zeroing is difficult because touching of the sensor with fingers has left body oils on it. When this happens, clean the sensor with a swab using a solvent such as isopropyl alcohol and be careful not to touch the sensor again.

FIM: Even relatively insensitive devices are at risk if the "antennas" are large enough and the **field** is strong enough. The possibility of **FIM** damage is why **fields** are a hazard to be avoided in accordance with the **Basic Rule.**

Grounding: In theory, if all materials in the **SSW** were **conductive** and grounded, as in the idealistic "**conductive** approach" (see "**Basic Rule**" above), all surfaces would be at the same (zero) potential, and **fields** and **discharges** would be nonexistent so no ESD damage could occur. In practice, however, **antistatic** and **static-dissipative** materials are much used because these materials are safer for people and **ESDS items** than are **conductors,** and these materials are grounded in various ways, e.g., by being held by a grounded operator or by lying on a grounded **antistatic** or **static-dissipative** workbench surface.

The **grounding** lug of the workbench itself must be connected through a resistor to an earth ground, e.g., a water pipe, as required in Model Specification 1 in Chapter 6. As stated in the same specification, some equipment, e.g., ovens, may be connected directly (hard grounded) to commercial power ground.

HBM: In the **HBM** (MIL-STD-883C, Method 3015, with Notice 7 released in Feb. 1988), the touch of a charged person's finger is simulated by discharging a 100-pF capacitor through a 1500-ohm resistor. But human capacitance reaches 250 pF, and body resistance can be as low as 100 ohms (Ref. 2–9). Or, Ref. 2–2 gives these extremes as up to 500 pF and down to

50 ohms. Furthermore, a charged metal tool might touch an **ESDS** lead or terminal (the machine model) so that the resistance would be less than 1 ohm. When a person discharges through a tool, the rise time may be much faster than with a finger because of the sharpness of the discharge tip (Ref. 2-2). Another problem with the **HBM** is that data vary from one simulator model to another, so that the need for a calibration procedure in MIL-STD-883C has been suggested (Ref. 2-10).

Ratings of sensitivity by this test are commonly called "V-zap levels," and the simulators may be termed "zappers."

Humidification: This ESD-control technique has been much maligned, for example in comparison to **ionization.** One reason for criticism is that **triboelectric charging** persists even at high relative humidity as shown in the following table from Ref. 2-5:

ACTIVITY	STATIC VOLTAGES	
	20% RH	80% RH
Walking across vinyl floor	12 kV	250 V
Walking across synthetic carpet	35 kV	1.5 kV
Arising from foam cushion	18 kV	1.5 kV
Picking up polyethylene bag	20 kV	600 V
Sliding styrene box on carpet	18 kV	1.5 kV
Removing Mylar tape from PC board	12 kV	1.5 kV
Shrinkable film on PC board	16 kV	3 kV
Triggering vacuum solder remover	8 kV	1 kV
Aerosol circuit freeze spray	15 kV	5 kV

However, the fact remains that **triboelectric charging** becomes especially troublesome below about 20 or 30% relative humidity, as shown by the high voltages attained at 20% RH in the above table. Incidentally, relative humidity values preferably should include an associated temperature. A temperature factor is involved in surface resistivity as we shall show in a future publication. In the above table, a temperature of about 70°F is assumed. Another problem at very low relative humidity, e.g., 10% at 70°F, is that the sweat layers on **antistatic** materials may work less effectively.

Several publications, e.g., Ref. 2-11, have mentioned the seasonal correlation of naturally occurring high humidity (usually in summer) and increased yields of **ESDS** products because of less ESD damage. Thus **humidification,** e.g., to 30 or 40% relative humidity, minimum, at 70°F, is surely desirable, but drawbacks include (1) expense of facilities for adding water to the air, (2) possible adverse effects such as delamination of polyimide circuit-board laminates or corrosion of metals if the humidity becomes too high, and (3) the psychological factor of false confidence inspired in operators and even engineers. If **humidification** can be done cost-effectively, unobtrusively, and without ill effects such as corrosion, do it. But keep in mind that ESD problems will not be removed but only lessened. No disciplines may be relaxed. As proof that ESD damage can occur at 40–50% relative humidity at 70°F, note the considerable charges, voltages, and **MOSFET** damage recorded in Papers No. 6 and 10 in the Appendix.

Insulator: See **"Nonconductor."**

Ionization: As we use this term, it refers not to the phenomenon in general but its application, in the form of air ionization, to ESD control.

Ionization is not a cure-all, as stressed in the following opening remarks we made in the **Ionization** Workshop at the 1988 EOS/ESD Symposium in Los Angeles:

> "**Ionization** for no valid, well-defined reason is useless and may actually be part of the ESD problem by creating charges on ungrounded **conductors;** that such charging can occur is indicated by GIDEP Alert H7-A-85-02, issued in 1985, on an electrically unbalanced AC ionizer. Hence, ionizer manufacturers have shown much concern about balancing, which in some models is now automatic.
>
> "In contrast to **humidification,** which reduces triboelectric charges by 'nipping them in the bud,' i.e., by inhibiting them in the first place, **ionization** can neutralize a charge only after it has formed. Thus, **ionization** is a process of correction rather than prevention.
>
> "However, **ionization** does have its place in ESD control when used judiciously. Local ionizers are helpful in tribocharging processes such as grit-blasting or peeling tape from a roll, and room **ionization** does control standing charges on common plastics, garments, etc. Room **ionization** is claimed to have an incidental 'air-cleaning' effect (repulsion of particles from surfaces or precipitation of particles away from work zones) in cleanrooms. However, particles adhering to surfaces, such as walls, by 'static cling' can be undesirably released by area **ionization** so that contamination is increased at least temporarily.
>
> "In conclusion, **ionization** is not a harmless vitamin pill but is strong medicine which has possibly dangerous side effects and should be used only when risks are carefully controlled and are outweighed by benefits."

Electrical ionizers have the possible problem of charging isolated **conductors** as mentioned above, whereas nuclear ionizers have a possible problem of polonium-210 contamination; one brand of nuclear ionizers was recalled by the Nuclear Regulatory Commission in February 1988 (Ref. 2–12). Polonium-210 is harmless unless inhaled or ingested, but the oral radiotoxicity is similar to that of plutonium—*if* the polonium-210 escapes from insoluble microcapsules that would be expected to pass harmlessly through the gastrointestinal tract (Ref. 2–13). The radiation hazard seems low, but it must be considered.

Slow-pulsed DC ionizers present a special problem, as explained in Example 21 of Chapter 5. We have reversed the recommendation of these ionizers which we made in Paper No. 3 in the Appendix, because when writing that paper we wrongly assumed that electrically isolated **conductors** would be grounded in all **SSWs**. In practice, **conductors** such as microscopes with rubber feet were not grounded and were observed to be throwing arcs to adjacent **conductors** when near a pulsed DC ionizer operating at 22 cycles/minute. Since AC ionizers do a similar job less dangerously, we advise using them instead of the slow-pulsed DC type. However, faster-pulsed DC ionizers may be acceptable; see Example 21 in Chapter 5. Note that even AC ionizers can charge electrically isolated **conductors** if the ionizers go out of balance. Hence we require autobalancing ionizers in paragraph 3.6.8 of Model Specification 1 in Chapter 6.

Beware of ozone generation as mentioned in paragraph 5.2.1 of Model Specification 1 in Chapter 6. The rule of thumb is that if ozone can be smelled its concentration is too high.

Ionizing blow-off nozzles are a special case. Gases when free of solids and liquids do not cause tribocharging (Ref. 2-14), and "many shop and dry nitrogen lines have been tested and no charging has been observed on aluminum and Teflon surfaces" (Ref. 2-15). Therefore, "installing ionizers on air lines because of charging fears is unwarranted" when dust is not of concern (Ref. 2-15). However, flying dust or other particles can cause **triboelectric charging,** so ionizing blow-off nozzles are a desirable precaution, and their use is suggested in DoD-HDBK-263. Ordinary air ionizes better than dry nitrogen, so a nitrogen nozzle can be configured to allow air to be aspirated into the gas stream ahead of the ionizing element; this feature appears in electrical designs but not in nuclear ones, so the electrical type should be used on nitrogen lines for best results.

Note that there are several possible causes of **static charge** formation in a blow-off operation. These include (1) **triboelectric charging** by flying dust or other particles or by sliding of displaced objects, (2) evaporation of nonaqueous solvents, (3) deformation of solids (piezoelectric effect), and (4) cryogenic charging, e.g., by freezing of water. In the last case, small splinters supposedly are ejected from the freezing drop, and they either carry off or leave an excess of electrons so that the remaining drop freezes to a charged ice particle.

In conclusion, **ionization** has its pros and cons. Pros are: (1) background **static charge** levels are reduced and (2) **ionization** is the most practical method of neutralizing charged **nonconductors;** indeed, ionizers are sometimes essential, as in solving the problems in Examples 25 and 34 of Chapter 5. The cons are: (1) dangerous space charges which charge ungrounded (electrically isolated) **conductors** are possible, (2) ozone may be generated above the OSHA limit, (3) the neutralization process (rate of charge decay by air ions impinging on a surface) is relatively slow so that ESD damage can occur in the course of **triboelectric charging** (the **field** created by pulling tape from a roll will light a neon bulb in the midst of a stream of ionized air), (4) work areas are incompletely covered by blowing ionized air because of shadowing by obstructions, and (5) **ionization** is ineffective when a **static charge** is hidden by **voltage suppression.** As an example of the latter, a charged sheet of plastic lying on a workbench surface is immune to neutralization by ionized air because the **field** is collapsed and cannot attract air ions of the opposite polarity. However, the **field** reappears and may cause damage by the **FIM** when the plastic is lifted from the surface. To put it another way, the plastic sheet lying on the surface makes an effective capacitor so that C is relatively high and V is very low in the equation $Q = CV$ (see "**Capacitance**" above).

In view of these advantages and disadvantages, we recommend using **ionization** only "when risks are carefully controlled and are outweighed by benefits," as we said above. Select ionizers from a major manufacturer with an established reputation, and insist on monitoring and autobalancing features to minimize the space charging hazard.

Latent Failure: This is one of many controversial subjects in the field of ESD control. **Latent failure** is accepted as being a real phenomenon (for example, Refs. 2-2, 2-16, and 2-17). But the question remains: how frequent are **latent failures**? The consensus of opinion by several leading ESD experts at the EOS/ESD Symposium in September 1988 was that these failures are indeed real but are very rare. Nevertheless, one **latent failure** in a high-reliability aerospace system, for example, might be catastrophic, so full ESD precautions must be taken even if the chance of **latent failure** is remote. The above comment on rarity pertains to total failures, but some device experts claim that malfunctions such as intermittency are not rarely but *usually* caused by ESD events in the manufacturing process. In critical applications such as life-support systems or computerized weaponry, such malfunctions, though short of complete failure, might themselves be disastrous. In conclusion, the possibility of **latent failure,** or at least latent malfunction, is a good reason for rigorous ESD control.

MOSFET: This has been the "white rat" in much of our research. It is appropriate for worst-case testing when the **ESDS** devices being protected are less sensitive than the **MOSFET,** as in our work. Other methods of recording voltage transients may be used, but **MOSFET** damage is graphic and convincing because an actual component is degraded. Endless **special tests** with **MOSFETs** can be devised; see Examples 10, 12, 18, 19, 21, 26, 27, 30, 31, 33, and 35 in Chapter 5.

Nonconductive, Nonconductor: DoD-HDBK-263 describes **nonconductors** as having a **surface resistivity** of 10^{14} ohms/square or more, but we consider the limit to be lower, namely 10^{12} ohms/square, because a **surface resistivity** below this level is needed for rapid bleed-off of **static charges** as discussed under "**Antistatic**" above. On a **nonconductor,** a **static charge,** true to its name, is static or motionless more than long enough (for only nanoseconds are needed) for its associated **field** to threaten **ESDS items** by the **FIM.**

Operator Disciplines: This subject is discussed in Chapter 4 on the **SSW,** and disciplines are incorporated in Model Specification 1 in Chapter 6. No matter how excellent the ESD-control methods and equipment, the operator is all-important, as the pilot of a plane is all-important, because the best equipment in the world can be misused; ESD damage can be done, or the plane can crash. See the discussion of the importance of operator skills in Reference Document 6 in Chapter 6.

Personnel Voltage Tester: In using this instrument, a charged person merely pushes a button to record his instantaneous voltage. For examples of test data, see Paper No. 10 in the Appendix. This is voltage on the skin, of course, not on garments.

Shielding: Even materials which are less than **conductive** provide partial **shielding** from **fields;** for example, an **antistatic** polyethylene bag gave 75% **shielding** in a **shielding/discharge** test (Table 9 of Paper No. 3 in the Appendix). However, **conductors** are required for a high level of **shielding.** We prefer a "buried" layer of at least 0.25 mil of aluminum foil (or the electrical equivalent) for bags or other containers for in-plant handling; see "**Faraday Cage**" above. For troubleshooting or solving unique ESD problem situations, **shielding** can be provided by various **conductive** barriers, e.g., a chain link fence protecting an **SSW** in Example 29 of Chapter 5 or a connector shell protecting **ESDS** pins in Example 30.

Like **grounding, ionization** (used only when necessary), or the use of **topical antistat, shielding** is a basic ESD-control technique.

Shunting: **Shunting** of **ESDS** leads or terminals does not assure complete ESD safety. For the best measure of safety, the consensus of opinion at present is that **antistatic** or **static-dissipative** shunt materials, e.g., foams, are insufficient, and **conductive** materials are required. Our test on black **conductive** (graphitic carbon-loaded) "noncorrosive" foam with low sulfur and chloride content showed it to be adequate for at least short-term high-humidity service, as was found for a solid shunt bar of **conductive** plastic in Example 14 of Chapter 5. However, long-term galvanic corrosion is possible as explained in that example, and the issue of galvanic corrosion must be considered in selecting and using any shunt material. See Table 1 of MIL-STD-889B and regard carbon-loaded plastic as "graphite."

Special Test: Numerous **special tests** appear in Chapter 5. The strategy is to start with conditions that are extremely worst-case, giving the highest confidence if the test is passed, and then, if the test is failed, back off to less exaggerated conditions that still give high confidence. An example is the replacement of a highly **ESDS coupon** design with a less-sensitive one, which still gave high confidence when it passed the test; see Summary point No. 5 of Paper No. 8 in the Appendix.

A **special test** can solve any ESD-control problem if the investigator is sufficiently ingenious. Remember that the object of testing is to reveal ESD hazards and then impart confidence in the means taken to solve them. The only danger of worst-case testing, besides revelation of unpleasant truths, is that an overkill solution might be impractical or expensive, but, if necessary, further tests at progressively milder conditions will give enough data for a cost-effective compromise between confidence and expense to be chosen. Overkill is desirable when cost-effective, as in the case of foil laminate bags that give far superior protection without costing more than see-through metallized bags (see Chapter 4) or a stone house built near a quarry.

SSP: We prefer a **Faraday-cage** package but allow a loophole, involving a **discharge test** (paragraph 4.3.3 in Model Specification 2 in Chapter 6), for suppliers to use nonfoil constructions. See Chapter 3 for a full discussion.

SSW: See Chapter 4 for a full discussion.

SSZ: The **SSZ** is the interior of the **SSP,** or a zone within the **SSW,** wherein the **Basic Rule** is never violated. See Chapters 3 and 4.

Standard Test: Examples of Government or industry **standard tests** are: the **HBM** test in MIL-STD-1686A, **surface resistivity** by ASTM D257 or DIN 53842, decay time by Method 4046 of FED-STD-101, resistance tests according to NFPA 99 (formerly NFPA 56A), and a test for **triboelectric charging** inside bags in Appendix G of EIA 541. (EIA 541 is an Electronics Industries Association standard.)

Our own **standard tests** include the **discharge test** in Model Specification 2 in Chapter 6, the **surface resistivity** test with a Voyager SRM-110 meter or equivalent (using our own criteria for classifying surfaces as **antistatic** or **nonconductive**), the wet-bulb/dry-bulb method for relative humidity, and the roller test, drag test, scuff test, and **walk test** for **static-limiting floor finishes** in Paper No. 6 in the Appendix.

Static Charge: For practical purposes, electrons can be imagined as "rubbing off" one surface to "contaminate" another, thus leaving the first surface positive and making the second one negative. Only if the **surface resistivity** falls below about 10^{12} ohms/square will the electrons cease to be static (stationary) and slide from the surface to give the rapid bleed-off that an **antistatic** surface needs to be useful in ESD control.

A **static charge** commonly results from **triboelectric charging,** but there are other sources, including freezing of water, evaporation of solvents, ion and electron beams, atomizing of liquids (spray charging), photoelectric charging, corona charging, and deformation of solids (piezoelectric effect of crystals).

Static-Dissipative: We use the standard definition from DoD-HDBK-263.

Static Field: See **"E Field"** above.

Static-Limiting Floor Finish: This subject is thoroughly discussed in Paper No. 6 in the Appendix.

A **static-limiting floor finish** is a backup or safety-net precaution like **humidification.** Another is room **ionization,** which we don't recommend except for cleanrooms when advantages, e.g., cleaning of the air, are demonstrable and risks minimal. Unlike these other methods, the special floor finish is inexpensive and quite unobtrusive. Tile floors must be "waxed" somehow, so why not use a **static-limiting floor finish**? Choose one that is durable and easily maintained by standard, simple procedures. **Triboelectric charging** of shoe soles is the property of interest, but **surface resistivity** is related to this *for a given finish* and is conveniently monitored; 10^{11} ohms/square by a Voyager SRM-110 meter is a typical allowable maximum before the floor coating must be restored. (*Note:* Among different finishes, the correlation between **triboelectric charging** and **surface resistivity** is imperfect; see Paper No. 6 in the Appendix.)

Remember that basic floor-finish properties, as well as ESD properties, are vital. For example, slip resistance must be retained throughout the wear cycle for employee safety.

Be careful to follow the floor-finish supplier's instructions for maintenance of the finish. For example, don't use diluted finish as a cleaner, because the resulting sticky deposit will collect dirt; use a special floor-finish cleaner, containing an antistat, which will restore the **antistatic** property, without stickiness, to the surface.

Surface Resistivity: Values obtained for this property depend on electrode configuration, pressure on the electrodes, roughness of the surface being measured, applied voltage, and other factors, so that methods of measurement are a deep subject (Ref. 2–18). However, for our purposes an approximate, easily made determination by a Voyager SRM-110 meter or equivalent is sufficient. The critical property is bleed-off time for **static charges,** and since we don't have an exact criterion for bleed-off time, which is arbitrary, we don't need precise values for the property, **surface resistivity,** that controls bleed-off time. As we said in a discussion of test methods (Ref. 2–1), "We need only to know if a feeble current will run across [a plastic surface]. So trying to split hairs on the exact resistance is not very productive thinking." In other words, an acceptable **surface resistivity** value, e.g., below 10^{12} ohms/square, is just an indication that enough current will flow across a surface so **static charges** can't accumulate.

One **surface resistivity** meter, such as the Voyager SRM-110 or SRM/RTG Meter, which measures resistance to ground also, is desirable for every assembly area where **ESDS items** are being handled. The meter will see much use in checking floor finish, questionable plastics, performance of **antistatic** materials when the relative humidity drops, etc.

Topical Antistat: **Topical antistat** treatment is almost a panacea for ridding the **SSW** of **nonconductors.** We say "almost" because **topical antistats** have their deficiencies: invisibility, fugitivity, and possible contamination.

Because of the invisibility of the antistat, treated surfaces must be "so labeled," as required in paragraph 3.6.10 of Model Specification 1 in Chapter 6. Fugitivity means that the antistat will disappear by vaporizing or wearing off at some unpredictable rate, so periodic checks must be made with a **field meter;** fortunately, this is easily done. For possible contamination effects, see Paper No. 9 in the Appendix. If corrosion by chloride ion is feared, e.g., when silicon wafers are being handled, use one of the "chloride-free" (actually, very low in chloride) **topical antistats** on the market.

Despite the above limitations, **topical antistats** are a good friend in ESD control, but don't overdo your reliance on them. Eliminate as many **nonconductors** as possible before antistat-treating the necessary ones remaining in the **SSW.**

Triboelectric Charging: The triboelectric series is only approximately valid; for example, an anomaly is noted in the complete version of Paper No. 6 condensed in the Appendix. However, for reference, here's a brief version including modern materials and deleting the irrelevant fur and sealing wax:

Positive End

Air
Hands
Glass
Hair
Quartz
Nylon
Wool
Silk
Aluminum
Paper
Cotton
Steel
Wood
Rubber

Gold
Polyester
Polyurethane
Polyethylene
Polypropylene
Vinyl
Aclar (polychlorotrifluoroethylene)
Silicon
Teflon
Negative End

In theory, when two substances from the list are rubbed together the one higher on the list becomes positively charged. Thus Aclar, which is a useful material when **triboelectric charging** (or "triboelectrification") is wanted, almost always takes a negative charge. Note that the same material, e.g., polyethylene, can charge itself so that one area of a sheet may be positive and another negative. Also, "circles" occur in which A is charged positively by B, B is charged positively by C, but C is charged positively by A instead of being charged negatively as it "should" be. Experimentally, **triboelectric charging** is notoriously erratic.

Triboelectric charging tests include the bag shaker method using quartz and Teflon disks to bracket most of the triboelectric series (Appendix G of EIA 541) and the roller test or drag test illustrated in Figs. 5 and 6 of Paper No. 6 in the Appendix.

Voltage Suppression: This is both a concept and a technique for ESD control. A good example of **voltage suppression** is the minimal **apparent charge** that can be observed on painted, grounded metals (allowed in the **SSW** as necessary **nonconductors** in paragraph 3.6.10 of Model Specification 1 in Chapter 6) or plastic-coated metals (also allowed in the same paragraph of Model Specification 1; note the vinyl-coated steel tote box in Paper No. 7 in the Appendix). Another example is the relatively low **apparent charge** on hair hanging close to the skin (Example 2 in Chapter 5).

Use **voltage suppression** whenever possible to "kill" **fields** and achieve compliance with the **CD Rule.**

Walk Test: This test is illustrated in Fig. 7 of Paper No. 6 in the Appendix. It may be augmented or replaced by measurements with a PVT-300 Personnel Voltage Tester as utilized in Paper No. 10 in the Appendix. However, when using the PVT-300 be sure to take at least 10 readings in order to catch, or partially catch, the voltage spikes occurring when the shoe leaves the floor and capacitance drops.

The **walk test** is a good example of a bottom-line practical and realistic test that directly measures the property of concern—voltage on people in this case. Laboratory tests tend to be more reproducible but less relevant to real-life use of materials.

Zapflash: This inexpensive and convenient continuity-checker can also be used to demonstrate various ESD phenomena such as voltage surges on a person caused by **triboelectric charging** when he shuffles his feet on a carpet. We recommend that every **SSW** have a **Zapflash** for verifying **grounding** as well as a **field meter** for use in complying with the **CD Rule.**

"Zapflash" is a trade name of Anderson Effects, Redlands, CA. Besides showing continuity up to about 7 megohms, the **Zapflash** can be used for checking batteries, lamp filaments, and fuses, for determining diode or DC voltage polarity, for revealing AC power leakage to grounded equipment cases or machinery (which could be a **DI** hazard), and for identifying the hot lead of an AC outlet.

Chapter 3

The Static-Safe Package (SSP)

By the **Basic Rule,** an **ESDS item** must always be in an **SSZ.** This **SSZ** will be in (1) an **SSP,** (2) an **SSW**, or (3) elsewhere under continuous operator supervision (paragraph 3.5.2 of Model Specification 1 in Chapter 6). An example of "elsewhere" is an open tote box being carried between **SSWs** by an operator whose wrist-strap cord is connected to an overhead grounded trolley.

In this chapter we're concerned with the **SSP,** which is defined in Chapter 2. The **SSP** must have an outer **Faraday cage** or at least pass a **discharge test** in accordance with paragraph 4.3.3 of Model Specification 2 in Chapter 6.

But how is the **SSZ** maintained inside the **SSP**? Remember that we're guarding from **fields** and **discharges** in accordance with the **Basic Rule.** Since the **Faraday-cage** wall of the **SSP** excludes external hazards, and there's no source of internal **discharges** except internal **fields,** the only internal hazard is the creation of **fields** by **triboelectric charging.** But this is a big "only," because **triboelectric charging** is a mysterious or at least erratic process affected by the least amount of surface contamination. For example, shoe soles charged the opposite of their original polarity when they became invisibly soiled, and sometimes a sheet of plastic, oddly enough, can be folded over and rubbed to charge itself, becoming positive in one spot and negative in another, perhaps because no two surfaces are quite identical (Ref. 2–14).

Thus, the problem of designing an **SSP** is attaining high confidence that **triboelectric charging** inside the package will be minimal. Approaches are: (1) running laboratory **triboelectric charging** tests, e.g., the test in Appendix G of EIA 541 or the roller test illustrated in Fig. 5 of Paper No. 6 in the Appendix, to screen candidate materials for an intimate wrap, (2) running a more realistic **special test** in which **coupons** with **MOSETs** are shaken inside the package as in Ref. 2–7, (3) mounting the packaged **ESDS item** so that it cannot slide within the package during shipment, and (4) protecting the item (but not completely, as mentioned under **"Shunting"** in Chapter

2) by **shunting** the leads or terminals. In approaches 1 and 2, the surface(s) of the item to be packaged should be tested in the **triboelectric charging** test; the EIA 541 bag-shaking test uses quartz and Teflon to bracket the triboelectric series (see **"Triboelectric Charging"** in Chapter 2) because they're at opposite ends of it, but, since the series is fallible, this method of scoping all materials may fail in practice.

At the present state of ESD-control knowledge and material development, we recommend using **antistatic** materials as the intimate wrap, keeping in mind the question of their permanence (Paper No. 2 in the Appendix) and taking no chances of contamination (Ref. 3–1) or corrosion (Paper No. 9 in the Appendix). If at all possible, secure the item so that it can't slide when the package is vibrated in shipment, and use the technique of **shunting.** Probably the worst hazard is static-generating foam such as polystyrene foam "peanuts." If such a material must be used, make very sure it's **antistatic** by a practical shaking test followed by a check with a **field meter.** Remember that the package may find itself in relatively dry air. Will the cushioning material remain **antistatic** at, say, 10% relative humidity at 70°F? Testing is necessary. In our experience, a good brand of **antistatic** bubble-wrap is a safer cushioning material than "peanuts." The bubble-wrap can't break open to expose fresh, **nonconductive** surfaces, and its resilience tends to hold packaged items in place. Use it whenever possible. Cardboard boxes with a buried layer of aluminum foil and with **antistatic** bubble-wrap inside are excellent packages.

Though we allow a variety of **shielding** materials to accommodate suppliers, as long as the **discharge test** in Model Specification 2 can be passed, we far prefer a **Faraday cage** (by our definition). In a hierarchy of tests (Paper No. 4 in the Appendix), a foil laminate bag was arbitrarily ranked as "good," while a bag construction with heavy foil (opaque, of course) or metal screen (partially transparent) was "excellent." Bags with see-through metallization were at best "fair." Significantly, two groups of investigators at major corporations arrived independently at the same conclusion that foil provides very effective **shielding** whereas see-through metallization, which is on the order of only 100 angstroms thick, is dubious. At the 1984 EOS/ESD Symposium, the Rockwell group said, "Only foil-containing laminate bags are recommended for **Faraday-cage** protection against worst-case **fields** and **discharges**" (conclusion 12 of Paper No. 3 in the Appendix), and the British Telecom group said, "Highly static-sensitive components should always be protected by metal-foil bags" (Ref. 3–2).

See-through metallized bags have become popular because of their partial transparency, and a **capacitance**-probe **(shielding/discharge)** test (in EIA 541) has been contrived by the bag manufacturers and accepted by naive users. The first problem with this test is that the voltage is too low (1000

V); the British Telecom tests (Ref. 3-2) used higher voltages, e.g., 1600 V, which can easily be encountered in practice. Furthermore, Ref. 3-2 shows that as the area of the electrodes falls the **shielding** effectiveness of see-through metallized bags decreases. This is in the direction of a spark **discharge,** and the second problem with the test is that spark **discharges,** which are not accounted for with the large flat electrodes used, burn off metallization and can damage **ESDS items** in the bag; a person can walk on a carpet at 20–30% relative humidity at 70°F and touch his finger to a see-through metallized bag to "blow" a **MOSFET** inside; see Fig. 2 and Table 3 of Paper No. 4 in the Appendix.

Admittedly, the **MOSFET** was configured in the worst possible way (leads connected to upper and lower "antennas"), so that the chance of ESD damage was maximized, but why take chances? We agree with Ref. 3-3 that "packaging cost differences of as much as one or two dollars for product protection in these areas (precision medical, communications, personal hazard, or defense applications) is a minor consideration. . . . EMI/RFI testing may also be appropriate in such cases." Yes! And foil or screen gives EMI/RFI protection along with protection from spark **discharges.** Furthermore, foil laminate bags cost little or no more than see-through metallized bags! Screen bags (see Paper No. 4 in the Appendix) do cost considerably more, but an extra "one or two dollars" is trivial when the product in the bag is worth $100,000 or its **latent failure** might cause a $10,000,000 catastrophe. Being penny-wise and pound-foolish in this case is not only risky but may be unethical where health or defense is involved. Note the suppression of data mentioned under **"Discharge Test"** in Chapter 2. Also, MIL-HDBK-773 (Ref. 3-4) insists on foil bags.

But we're tired of being a voice crying in the wilderness. Let the world go its way and make believe that a wispy 100 angstroms of metal, which is not an EMI/RFI shield and is easily evaporated by a spark from a finger, is a "Faraday cage"!

For recommended **Faraday-cage** tote box constructions, see Paper No. 7 in the Appendix; the **shielding** layer in these is foil as a minimum.

Here's a final word on corrosion and contamination (the subject of Paper No. 10 in the Appendix). We advise running an accelerated **special test** with your **ESDS item** inside of, or against the surface of, the proposed intimate wrap, as a test was done with a connector and the shunt bar proposed for it in Example 14 of Chapter 5. Generally suitable high-humidity conditions are MIL-STD-202, Method 106, with or without the vibration or freeze cycles as appropriate, for one month. This rigorous accelerated test probably will disclose long-term problems, but we suggest an accompanying real-time test at slightly elevated temperature, e.g., 100°F, with samples being checked every few months. The results of the real-time test will give you a good idea of what is happening to your product in storage. Where corrosion is concerned, "never assume anything."

For suppliers of ESD-control packaging materials, see buyers' guides such as the *EOS/ESD Technology 1989 Buyers' Guide,* Miller Freeman Publications, Inc., San Francisco, CA, or the similar guide from *Evaluation Engineering* magazine, Nokomis, FL.

Chapter 4

The Static-Safe Workstation (SSW)

How the SSW Holds the SSZ

As mentioned at the start of Chapter 3, an **ESDS item** must always be in an **SSZ.** This chapter is concerned with the means of maintaining an **SSZ** inside an **SSW.**

First, note that the **SSZ** may be irregularly shaped and may shift position within the **SSW.** For example, if an ionized air blower is being used, the **SSZ** will end 12 inches from the corona-discharge points (paragraph 3.6.8 of Model Specification 1 in Chapter 6), and when the ionizer is shifted on the workbench the **SSZ** will also shift. The operator is responsible for maintaining the **SSZ** and knowing its boundaries at all times. We stress again that his or her skill, not the selection of materials and equipment (though these must be adequate), is the most important element of ESD control.

Since the **SSW** is a volume in space, it is not just a working surface. Floors, ceilings, and walls may or may not be included, and the volume in space may contain only part of a workbench or no workbench at all; for example, an oven might be an **SSW.** Whatever the **SSW** is, it must be "designated" (see definition in Chapter 2), which means clearly marked, e.g., with strips of yellow tape.

Can an **SSW** have any **fields** at all? Yes, but they are known and controlled so that they're always harmlessly weak within the **SSZ** as required by the **CD Rule.** If no **fields** whatsoever were permitted within the **SSW,** the operator himself or herself would be excluded because hair and clothing "radiate" **fields.** These **fields** must, of course, be sufficiently attenuated by distance at the point where the operator is handling the **ESDS item;** if the attenuation is insufficient, so that the **CD Rule** is violated by too short a distance from **apparent charges** on hair or clothing, corrective actions are required, e.g., tying the hair closer to the skin or changing the clothing, for example from wool to cotton.

Thus the operator's head and body are outside the **SSZ** while his or her hands are inside it. Why not try to force the whole operator, chair and all, to meet **SSZ** requirements? The reason is cost-effectiveness. We're using operator skills to minimize purchases of **antistatic** smocks and **conductive** chairs as well as to avoid troublesome restrictions such as the wearing of hair nets, and experience has shown that this cost-effective approach is feasible. The operator need only remember that **ESDS items** must be "kept at arm's length" and not be needlessly brought near hair, clothing, or chair upholstery.

Similarly, walls and floors with **fields** harmless by the **CD Rule** may be inside the **SSW.** In fact, the **SSW** should be considerably larger than the **SSZ** so that sources of **fields** near the **SSZ** are controlled. Consider Example 29 in Chapter 5. In this case the boundary of the **SSZ** was 18 inches from a chain link fence which provided **shielding** from **fields** on passers-by. Should the boundary of the **SSW** be the same as that of the **SSZ,** or should the **SSW** extend all the way to the fence? It should extend to the fence so that the 18-inch "buffer zone" along the **SSZ** is subject to **SSW** requirements and new **field** sources won't be added. If the **SSW** ended 18 inches from the fence, common polyethylene packaging film carrying a high **apparent charge,** e.g., −10,000 V, might be "legally" heaped between the fence and the **SSW.** In theory, the operator would note any violation of the **CD Rule** by using his or her **field meter** and would remove the polyethylene, but if the **SSW** had extended all the way to the fence the plastic wouldn't have been left lying there (assuming compliance with **SSW** rules) in the first place.

Though the language was unclear, the old "one-meter rule" of DOD-STD-1686, now superseded by MIL-STD-1686A, seemed intended to provide such a buffer zone. The trouble with this rule was that it was arbitrary, and one meter might not be a safe distance, by the **CD Rule,** from a high **apparent charge.** Since we have the quantitative **CD Rule,** the extent of the buffer zone between the working surface and the boundary of the **SSW** is optional; it may be nil where a workbench touches a wall but may be a few feet from the other end of the bench as marked by a yellow line on the floor. In this case, common **nonconductive** plastic couldn't "legally" be left lying on the floor near the bench. Without wasting expensive space, of course, such buffer zones should be maximized.

Certification of SSWs

See Reference Document 2 attached to model Specification 1 in Chapter 6.

Workbenches: Design and Grounding

A static-safe workbench should (1) hold no **apparent charge** over 50 V (when **ESDS items** sensitive to no less than 100 V are being handled, as assumed in this book), (2) allow charges to drain from charged objects placed on the work surface (unless the charges are on **nonconductors** and are thus truly **static charges** that can't flow), and (3) have appropriate electrical supply and ESD **grounding** facilities including a **grounding** lug.

Requirement 1 is satisfied by constructing the bench of painted steel which is ESD-grounded. A painted surface is **nonconductive** but shows a negligible **apparent charge** because of **voltage suppression** by the steel; hence "painted, grounded metal" is one of the "necessary **nonconductors**" allowed in paragraph 3.6.10 of Model Specification 1 in Chapter 6.

Requirement 2 is met by selecting a work surface that is **antistatic** or **static-dissipative** by a **surface resistivity** measurement (Ref. 4–1) and has a surface-to-ground resistance (Ref. 4–1) of 1×10^{11} ohms, maximum. A **conductive** surface is undesirable because it can be hazardous to workers, by carrying a lethal current, and also to **ESDS items,** by allowing rapid **discharges** (see **"CDM"** in Chapter 2). The effects of wear, chemicals, staining, and heat on the work surface (Ref. 4–1) must also be considered, as well as possible corrosion of **ESDS items** by impurities from the work surface. Corrosion is a treacherous problem because it takes time to manifest itself. If silicon wafers will rest on the surface, consider a test for extractable chloride ion in laminates as described in Example 17 of Chapter 5.

The importance of drain time is another controversial subject. See Ref. 4–2 for a discussion of the danger of **voltage suppression** hiding a **field** when a charged object such as a tote box rests on a surface so that the field returns to possibly cause ESD damage when the object is lifted. Paper No. 7 in the Appendix makes the point that contact area between object and surface complicates the issue; a warped rigid **conductive** tote box with three small contact areas will drain far more slowly than a similar box with a flat bottom giving a relatively large contact area. In practice, a grounded operator usually touches a box or package before reaching into it, so a **conductive** work surface, with its hazard to people and **ESDS items,** shouldn't be used just to reduce drain time. We would rather err in the direction of high drain time than of too-rapid **discharge** to the surface that might cause **CDM** damage.

Requirement 3 involves **grounding** the bench to a water pipe, building structural steel, or wet ground well as required in paragraph 3.6.5.2.1 of Model Specification 1 in Chapter 6. A resistor is needed between the bench-top laminate and ground because direct **grounding** of the **conductive** inter-layer of laminates is unsafe (Ref. 4–3). If high voltages are being used and the utmost personnel protection is needed, use ground fault circuit interruptors. These guard against the remote possibility of an electrically overstressed resistor being carbonized so that it falls below a safe resistance (Ref. 4–4).

Incidentally, a **Zapflash** is useful for locating AC power leakage onto cabinets, etc.; touch the probe to the surface and the bulb will light if there is leakage. Subsequent testing with a voltmeter will show the voltage.

For details on **grounding** of workbenches, see Model Specification 1 in Chapter 6.

Humidification, Ionization, Static-Limiting Floor Finish

These subjects are discussed in Chapter 2. In brief, use **humidification** if your ambient air is often dry, if the expense of **humidification** equipment is bearable, and if there are no harmful side effects such as corrosion; also beware of false confidence inspired in personnel. Use **ionization** only when there is no alternative, as sometimes happens; Examples 24 and 25 in Chapter 5 are cases where **triboelectric charging** was unavoidable and only **ionization** could mitigate it. Make the most of **static-limiting floor finish,** which is a very economical "safety net" as seen in the cost calculation in Table 1 of Paper No. 6 in the Appendix.

Materials

Allow "necessary **nonconductors**" as listed in paragraph 3.6.10 of Model Specification 1 in Chapter 6. **Nonconductors** are either ignored because their **apparent charge** is low and they are safely far from the **ESDS item,** in accordance with the **CD Rule,** or they are controlled by treating with **topical antistat** or by having their **fields** attenuated by **shielding,** as by the chain link fence in Example 29 of Chapter 5, or by **voltage suppression,** as by keeping hair close to the skin in Example 2 of Chapter 5.

Allow **conductive** materials only when no **antistatic** or **static-dissipative** alternative is convenient. (*Note:* As of this writing, the more convenient EIA designation of "dissipative" is coming into use to replace these two classifications with only one. This is a sensible change in nomenclature, because the difference between **antistatic** and **static-dissipative** is arbitrary and is seldom a useful distinction.) For categories of materials and criteria for their selection, see Reference Documents 4 and 5 appended to Model Specification 1 in Chapter 6.

When selecting materials, remember that **standard tests** may not be sufficient! Laboratory tests often give misleading results because real-life conditions are not simulated. Hence, **special tests** with realistic configurations are often needed. An example of an unrealistic test is the **shielding/discharge** or pulse test in EIA 541, which ignores the realistic possibility of a spark **discharge** as discussed in Chapter 3.

Another, far more serious example of unrealistic testing involves a type of aerospace wire. The insulation on this wire has three different properties from convenional wire insulations: (1) possible susceptibility to hydrolysis because it's a condensation instead of an addition polymer, (2) unusual stiffness because the insulation is high-modulus and must be wrapped on the conductor as tape, and (3) high aromaticity (high carbon content in the molecule).

If realistic testing had been done before the wire was installed in airplanes and space vehicles, the following would have happened: (1) exposure to warm water and high-pH cleaning solutions, simulating wheel-well conditions in Navy planes, would have caused hydrolytic breakdown, so the wire would have been restricted from wheel-well applications, (2) vibration tests with simulated airframe assemblies would have shown chafing and shorting when the stiff wires rubbed together (as opposed to no chafing with conventional wire whose soft insulation makes it limp), so installation would have been done with this problem in mind, and (3) flashover and arc propagation tests would have shown that an arc propagated even in vacuum because of carbonization of the highly aromatic polymer, and care would have been taken to prevent shorting that might start an arc, or the wire would have been excluded from use in applications where the current was sufficient to sustain an arc.

What really happened was that applications went forward based on standard, unrealistic lab tests. For example, laboratory abrasion tests gave excellent results because they did not include the chafing situation where stiff wires rubbed together, and **standard tests** for electrical properties failed to predict the arcing problem. Thus untold millions of feet of the wire have been installed, and all three problems described above have reared their ugly heads in the course of a few decades (disasters can be slow in coming). The wire is now "in big trouble" because realistic tests weren't run long ago. But enough. A word to the wise is sufficient.

Equipment

Model Specification 1 in Chapter 6 requires **grounding** of all **conductors,** which is not always easily done. For example, a stainless-steel oven shelf is grounded directly, but the blade of a screwdriver is grounded to the operator's hand only through the "sweat layer" of **topical antistat** applied to the **nonconductive** handle. Microscopes present a special problem requiring judgment in bending the **grounding** rules; metal eyepiece housings, for example, have small **capacitance** and are not expected to touch an **ESDS item,** so they can be ungrounded. However, we recommend **grounding** the major sections of the microscope even if wires must be added to link them. Use the **Zapflash** to check continuity.

Soldering iron tips must be grounded so that there is a potential of no greater than 2 mV between the hot tip and the grounded **ESDS item** being soldered, if DoD-STD-2000-1B (paragraph 4.11.3) is being followed. Also, that paragraph requires less than 20 ohms resistance between tip and item.

Electrical equipment must be checked for **E fields,** of course. Remember that space charging can be a problem with ionizers, as discussed under "**Ionization**" in Chapter 2, and don't let ionizers become part of the ESD problem they're supposed to solve.

Treat plastic parts of equipment, e.g., housings, with **topical antistat** unless the plastic is a coating that is relatively thin (less than 20 mils) and the metal is thick enough (over 60 mils) to provide effective **voltage suppression** (paragraph 3.6.10(2) in Model Specification 1 in Chapter 6). Beware of stresscracking polycarbonate (Paper No. 9 in the Appendix).

In special cases, beware of *H* fields (magnetic fields). DoD-STD-2000-1B, paragraph 4.11.3, requires that *H* fields of soldering irons, soldering machines, and associated processing equipment be less than 2 gauss measured at any surface of an item being processed.

Continuous wrist-strap monitors (see Papers No. 5 and 10 in the Appendix) are a must; an example of a procurement document for these is Reference Document 3 appended to Model Specification 1 in Chapter 6. Other essential pieces of equipment are a **field meter** and a **Zapflash.**

For categories of equipment and criteria for their selection, see Reference Documents 4 and 5 appended to Model Specification 1 in Chapter 6.

General Design, Construction, and Maintenance of ESD-Protected Areas

See Reference Document 1 appended to Model Specification 1 in Chapter 6.

Operator Disciplines, Training, and Certification

In general, operators must maintain the **SSZ** and always know its boundaries within the **SSW. ESDS items** should never be brought unnecessarily near or touched to any surface, nor should an operator fidget or shuffle his or her feet unnecessarily. In other words, though the best materials and equipment have been selected and the **ARTG** is being met by using a **continuous wrist-strap monitor,** don't look for trouble. See the "Conclusions" section of Paper No. 3 in the Appendix. Operator disciplines are included as requirements in Model Specification 1, paragraph 3.6.12, in Chapter 6, and training and certification of operators are covered in Reference Document 6 appended to that specification.

Chapter 5

ESD Troubleshooting; Illustrative Examples

Troubleshooting, by which we mean the solving of unique problems as they arise, is done using one or more of three approaches (**analysis, standard test,** or **special test**) with the aid of other real or conceptual tools from Chapter 2. In the following thirty-five actual case histories, we describe each problem and state which of the above three approaches was used. Then we present the solution in enough detail to illustrate the reasoning process.

In principle, any novel ESD control situation can be handled, as were these examples, with the A-to-Z arsenal of ideas and techniques of Chapter 2. Solutions to problems then can be incorporated into the handling specification as illustrated in paragraphs 3.6.12.6 and 3.6.12.11 of Model Specification 1 in Chapter 6.

1. Arm Hair

Problem: Reference 5–1 suggested that arm hair might be an ESD hazard, with charges up to 900 V having been measured. We suspected that such voltages are atypical. What new **operator disciplines,** if any, should be introduced to control **fields** from arm hair?

Solution: The approach was a **special test.** Polyester garment sleeves or sheets of common **nonconductive** polyethylene were stroked across the upper surface of operators' forearms, and the **apparent charge** was read. Because of **voltage suppression** by the skin, in no case did the reading exceed 300 V, which represents a harmless **field** for **ESDS items** sensitive to 100 V (Fig. 3 of Paper No. 3 in the Appendix). Therefore, no new **operator disciplines** were introduced. However, operators were warned to keep the potential hazard in mind and run checks with their **field meters** on especially dry days. Frequent **field-meter** checks of the whole **SSW**, to enforce the **CD Rule,** are of course a vital part of our program.

2. Head Hair

Problem: **Apparent charges** of up to +10,000 V were observed on the hair of female operators; by the **CD Rule,** the safe distance from 10,000 V is 56 inches (paragraph 3.6.11 of Model specification 1 in Chapter 6). What new **operator disciplines** should be introduced?

Solution: The approach was a **special test. Apparent charges** of up to +10,000 V were found with hair hanging 6 to 8 inches from the skin, but when the hair was 2 inches from the skin the maximum **apparent charge** was only +300 V, which is a harmless level as noted in Example 1 above, because of **voltage suppression.** Therefore, a new **operator discipline** was introduced: hair must be kept within 2 inches of the skin by tying back as required; this requirement appears in paragraph 3.6.12.6 of Model Specification 1 in Chapter 6.

3. Operator Clothing

Problem: What precautions should be taken regarding the **static charges** on clothing? Should special **antistatic** or other type of ESD-control smocks be required?

Solution: The approach was a **special test. Apparent charges** were measured on shirts, blouses, and sleeves being worn by operators, and a few months' worth of data showed that **fields** were sufficiently weak in the vicinity of the operator's hands, where the **ESDS item** was, to meet the **CD Rule** as long as sleeves were short or rolled to above the elbow. Therefore, street clothes are allowed if the sleeves are short or rolled; with long, unrolled sleeves, ESD-protective smocks are required (paragraph 3.6.12.10 of Model Specification 1 in Chapter 6). Smocks, though optional, may raise ESD awareness by being "uniforms" and give factories a more businesslike appearance. However, these marginal advantages must be weighed against the initial cost of the smocks as well as the cleaning costs, which may be considerable. A compromise is to use rolled-up sleeves in engineering and test labs and smocks in assembly areas.

4. Identification Badges

Problem: **Apparent charges** of up to −2000 V were observed on plastic identification badges (Table 4 of Paper No. 3 in the Appendix). What precautions should be taken?

Solution: The **apparent charges** of up to −2000 V were observed when the badge was held away from a surface; in fact, by **triboelectric charging** of badges with polychlorotrifluoroethylene (Aclar, by Allied-Signal), the apparent charge could be pushed as high as +10,000 V. However, in **special**

tests, when the charged badge was held close to the skin, as a badge is worn in practice, the **apparent charge** dropped to 500 V or less because of **voltage suppression.** The **CD Rule** allows a distance of 12 inches from 500 V (paragraph 3.6.11 of Model Specification 1 in Chapter 6), and **ESDS items** are seldom brought much closer than this to a badge. Furthermore, badges are not highly charged in practice because they're not rubbed as in our worst-case tests. The conclusion is that badges are not a practical hazard, but if there were any doubt they could be easily treated with a **topical antistat.**

5. Vinyl Chair

Problem: An **apparent charge** of +500 V was noted on a vinyl-upholstered chair (Table 4 of Paper No. 3 in the Appendix), and higher voltages could be produced in contrived **triboelectric charging** tests. Therefore, should ordinary chairs be allowed in the **SSW,** or are **conductive** chairs necessary?

Solution: The approach was **analysis.** First, the operator's sitting in the chair attenuates the chair's **field** by **voltage suppression.** Second, the bench top provides partial **shielding.** Third, **field-meter** checks in assembly areas have always confirmed compliance with the **CD Rule** in regard to chairs. The conclusion is that ordinary chairs are acceptable (paragraph 3.6.10 of Model Specification 1 in Chapter 6), but an **operator discipline** requires that **ESDS items** shall never be held beyond the front edge of the work surface (paragraph 3.6.12.2 of the same specification).

6. Paper Forms and Labels

Problem: Should paper forms and labels, used to document manufacturing operations, be allowed in the **SSW?**

Solution: The approaches were **analysis** and **standard test.** DoD-HDBK-263, paragraph 7.1.2.3, classifies paper products as **antistatic** materials. Also, we checked some of the subject forms and found a **surface resistivity** of 10^9 ohms/square at approx. 50% relative humidity and 72°F. This value is at the low end of the **antistatic** range, and experience has shown that unglazed paper products are not an ESD hazard except in high-speed processing such as printing. The conclusion is that paper forms and labels, if unglazed, are allowable in the **SSW.**

7. Charged Walls

Problem: Do charged walls present a **field** hazard when they bound an **SSW,** e.g., when a workbench is against a wall?

Solution: The approach was a **special test. Triboelectric charging** of walls with garment fabric or Aclar film gave less than 1000 V **apparent charge** because **fields** tend to collapse on large surfaces **(voltage suppression)**. At 1000 V, the safe distance in inches by the **CD Rule** = $(1000)^{0.5}/1.8 = 18$ (or use the convenient table in paragraph 3.6.11 of Model Specification 1 in Chapter 6), and in practice **ESDS items** are never nearer than this to walls in our operations. Therefore, walls are judged not a practical hazard. If they were, they would be treated with **topical antistat** in accordance with paragraph 3.6.10 of the same specification; note that the **topical antistat** must be renewed at least every three months.

8. Charged Windows

Problem: Under dry conditions (relative humidity approximately 20% at 70°F), an **apparent charge** of 5000 V was noted on the glass of a window. **Triboelectric charging** is not caused by pure air or any other gas (Ref. 2-14), so the charging was attributed either to electrically unbalanced **air** or to dust carried by the warm desert wind which also caused the low humidity. How should the hazard of charged windows be controlled?

Solution: The approach was **analysis.** By the **CD Rule,** the safe distance from the window in inches = $(5000)^{0.5}/1.8 = 39$, so **SSW**s were arranged so that their boundaries were at least 39 inches from windows. Otherwise, a nonsmearing **topical antistat** could have been applied to the glass, with **field-meter** checks to make sure the charge was sufficiently lowered and remained in control.

9. Cathode Ray Tubes (CRTs)

Problem: CRTs (computer or oscilloscope screens) have a considerable positive charge as needed for the electron beam to "paint pictures" with the phosphors on the inside of the glass. Since CRTs are often found inside **SSW**s today, how can this **field** hazard be controlled?

Solution: The approach was a **special test.** The **apparent charge** was found to be about +2000 V with an unshielded tube but less than +300 V with a grounded **conductive** mesh screen (Paper No. 8 in the Appendix). The screen was a commercial CRT accessory which reduced glare as well as providing **shielding.**

10. Connector Dust Cover

Problem: Was a **nonconductive** polyethylene dust cover suitable for a connector on a cable assembly leading to **ESDS items** inside an electronic box?

Solution: The approach was a **special test** as well as a **standard test** (the **walk test** depicted in Fig. 7 of Paper No. 6 in the Appendix). A person charged himself to −10,000 V (measured by the **walk test** method) by shuffling on a carpet. Then his finger touched the dust cover over the connector face just above a pin which was connected to the substrate-case lead of a **MOSFET** at the other end of the 3-foot cable, the gate lead of the **MOSFET** being grounded. When measured in the usual way, the **MOSFET** was found to be damaged (either shorted or $V_{GS(TH)}$ shifted 0.04 V or more). Another **MOSFET** was installed and was damaged when the charged person's finger was brought to within 1 inch of the dust cover. However, when the cable, which was an unshielded wire bundle, was covered with a grounded metal shield, five **MOSFETs** in a row survived the test in which the charged finger touched the dust cover; then another five survived with the cover removed. When the dust cover was replaced by aluminum foil wrapping the connector face, but the cable was unshielded, a **MOSFET** was damaged by the charged finger being touched to the foil.

These tests show that ESD-protection of the connector face was irrelevant. The problem was that the **field** from the charged person's hand was being "picked up" by the unshielded wire to the **MOSFET** so that this device was damaged by the **FIM**; the dust cover could be charged to 10,000 V by **triboelectric charging** at low relative humidity (10–15% at 70°F), so it itself might have caused damage by the **FIM** with the cable unshielded.

The conclusion was that when the cable was shielded the dust cover could be any material, even ordinary **nonconductive** plastic. However, good ESD-control practice demands that an **antistatic** or **conductive** plastic cover, either of which is commercially available, be used. (*Note:* A **static-dissipative** cover also would be fine; this is an example of the uselessness of distinguishing between **antistatic** and **static-dissipative**, which are suitably lumped together as "dissipative" as in EIA practice.) The most important result of the test was the demonstration of the need for **shielding** the wire bundle of the cable.

Note that these conclusions pertain *only* to the particular connector-cable configuration tested. For example, if a sensitive pin were less deeply recessed and were closer to the dust cover, a charged finger might "zap" the pin through **antistatic** or even **conductive** plastic; then a true **Faraday-cage** cover (buried foil or better) would be needed.

11. Brushes

Problem: The suitability of four types of brushes for use in scrubbing **ESDS** modules in **SSWs** was to be determined.

Solution: The approach was **analysis, special test,** and **standard test** (for

surface resistivity). The first brush had a varnished wooden handle, a phenolic band (seemingly wood-flour filled) holding the bristles, and natural (horsehair) bristles. AT 65% relative humidity and 71°F, the **surface resistivity** of the handle was 2×10^{12} ohms/square at 500 V and of the phenolic band was 4×10^{10} ohms/square. The bristles, being natural rather than synthetic, were approved by **analysis** in accordance with DoD-HDBK-263. The conclusion was that the only part of the brush that was a problem was the handle, which had to be treated with **topical antistat** and so labeled (as an insulated tool handle, paragraph 3.6.10(9) of Model Specification 1 in Chapter 6).

The second brush had an unfinished wooden handle and a metal band holding the horsehair bristles, while the third brush was similar but with hog bristles. The wood was found to be **antistatic,** and the bristles were approved by **analysis** as for the first brush.

The fourth brush was a novel design created to solve a problem. The steel handle of an acid brush was scraping modules, so a one-inch section of **nonconductive** polyolefin heat-shrink tubing was shrunk over the handle to cover the first eighth-inch of bristles (natural) and prevent the steel from touching the surface being scrubbed. But would the polyolefin hold an unacceptable **field? Triboelectric charging** of the detached polyolefin gave −390 V **apparent charge,** which would require a safe distance in inches, by the **CD Rule,** of $(390)^{0.5}/1.8 = 11$ between the polyolefin and an **ESDS item.** However, when the polyolefin was shrunk in place onto the steel handle, which was held by a grounded person, the **apparent charge** was reduced by **voltage suppression** to −110 V, which is acceptable on **nonconductors** (Fig. 3 of Paper No. 3 in the Appendix). The conclusion was that the brush was acceptable for use without antistat-treating the polyolefin, which in essence falls under the plastic-coated metals provision of paragraph 3.6.10(2) of Model Specification 1 in Chapter 6.

12. Screwdriver Handle

Problem: Could the plastic handle of a screwdriver be a **field** hazard in the **SSW**?

Solution: The approach was to use **special tests.** The screwdriver in question was 9 inches long overall with a cellulose acetate butyrate plastic handle 3¼ in. long and 1 in. in diameter. At approximately 50% relative humidity and 70°F, the handle was stroked with polyester fabric to give an **apparent charge** of +1700 V. **MOSFET**s were damaged by the **field** (see Table 4 of Paper No. 3 in the Appendix), and the **CD Rule** calculation gives the safe distance in inches as $(1700)^{0.5}/1.8 = 23$. The conclusion is that the handle indeed could be a hazard, so it was treated with **topical antistat** and so labeled. This and similar tests led to the provisions of paragraph 3.6.10 of Model Specification 1 in the Appendix.

13. Smocks

Problem: Were three different designs of **antistatic** or **static-dissipative** smocks suitable for use inside **SSWs**?

Solution: The approach was a **special test.** The first smock, containing 1% stainless-steel fibers, was tested by clamping the cloth between two sets of viselike electrodes; at 10–1000 V the **surface resistivity** by a megohmmeter was 2–4 × 10^{10} ohms/square. Alternatively, the resistance was measured between dammed-in pools of mercury to give 2 × 10^{10} ohms/square at 10 V.

The second smock had black **conductive** threads in a crisscross grid and measured 6 × 10^5 ohms/square **(static-dissipative)** by the clamp method. A third smock had black **conductive** threads 4 mm apart and running all the same way so that the result of the clamp test was 8 × 10^7 ohms/square with the black threads perpendicular to the electrodes but greater than 10^{13} ohms/square with the black threads parallel to the electrodes.

The conclusion was that the first two smocks were suitable in terms of **surface resistivity,** but the third smock was doubtful. Note that the use of smocks in general was judged to be optional in Example 3 above. Before the first design could be used, the possibility of its shedding stainless-steel fibers would have to be checked, because these might cause shorts in **ESDS items,** e.g., between adjacent circuit lines on printed wiring boards. GIDEP Alert D5-A-84-01 (April 25, 1984) describes just such an occurrence.

14. Shunt Bar

Problem: Would a black **conductive** plastic shunt bar cause corrosion of gold-plated Kovar connector pins?

Solution: The approach was a **special test.** An edge connector was cut in half, and one half was fitted with the shunt bar while the other half served as a control. The pins were cleaned by vapor-degreasing with freon TES and washing with deionized water before starting the test. After 31 days under high-humidity, elevated-temperature conditions (MIL-STD-202, Method 106, without freeze or vibration cycles), there was no sign of corrosion and no difference in appearance of the shunted versus control pins. The conclusion was that the shunt bar was suitable for general use at moderate relative humidity. Gold and graphitic carbon are both "noble" (MIL-STD-889), and coupling them doesn't promote galvanic corrosion, but any flaws or pinholes in the gold plating could allow eventual attack on the underlying basis metal during prolonged high-humidity storage.

15. Cleanroom Gloves

Problem: Was a particular brand of **antistatic** disposable vinyl gloves suitable for use in **SSW**s in cleanrooms? Could **nonconductive** latex gloves be used?

Solution: The approach was **analysis** and a **special test.** The supplier's literature claimed that a charge of 5000 V substantially drained from the **antistatic** gloves in less than 2 seconds when tested by FED-STD-101, Method 4046, and this rapid bleed-off of charge, along with a reduced propensity for **triboelectric charging,** is of course the basis for using **antistatic** materials in ESD control. A test with a **surface resistivity** meter verified the claimed **antistatic** rating. Therefore, these gloves were approved.

In a **special test** on the latex gloves, it was found that **voltage suppression** limits their **apparent charge** to under 300 V, which is a safe level on **nonconductors** (Fig. 3 of Paper No. 3 in the Appendix) when they are being worn. Therefore, these gloves are allowed as necessary **nonconductors** in paragraph 3.6.10(3) of Model Specification 1 in Chapter 6. They are used when antistats might rub off and cause contamination, as of silicon wafers. Precautions in using these gloves are: (1) they must be put on and taken off 5 feet from the boundary of the **SSW,** (2) they must fit tightly for **voltage suppression** to work, (3) they should not touch **ESDS** leads because some of the charge on the latex, though truly a **static charge,** might possibly be transferred, and (4) they should not be needlessly rubbed on any surface because the less they are triboelectrically charged the better, even though the voltage is suppressed. Furthermore, it must be realized that charges on items being handled cannot drain off onto the fingers when these nonconductive gloves are being worn.

16. Cloth Wrist-Strap Bands with Metal Fibers

Problem: Could these bands shed metal fibers which might cause shorts in **ESDS items** as reported for cloth with stainless-steel fibers in GIDEP Alert D5-A-84-01 (see Example 13, above)?

Solution: The approach was a **special test.** Three brands of stretchable cloth bands with stainless-steel fibers were tested by expanding the band and twisting it to its complete capacity 100 times and then rinsing it with clean freon and filtering to catch any shed particles or fibers for microscopic examination. No metallic fibers or particles above 25 microns in length were found for any of the bands. The conclusion was that the bands were acceptable.

17. Workbench Laminates

Problem: Were certain benchtop laminates corrosive toward **ESDS items**, e.g., silicon wafers, lying on the surface?

Solution: The approach was to use **special tests.** First, a clean copper panel was pressed against the surface of a sample of laminate and exposed to the elevated humidity and temperature conditions of MIL-STD-202, Method 106, without freeze or vibration cycles, for 1 week. The copper surface was iridescent purple because of oxide, and small amounts of green corrosion products, possibly copper chloride, were detected by observation at 8×. Next, using the same brand of laminate, a ring of room-temperature-vulcanizing silicon rubber served to dam a pool of deionized water, covered to prevent evaporation, for 6 days. Titration (Ref. 5-2) showed 0.26 mg chloride/in.^{2} of surface versus negligible chloride for a thermoplastic surface used as a control. Five other brands of laminate showed 0.00 mg chloride/in.^{2}, one brand showed 0.56, and another showed 1.1.

The conclusion was that five of the eight brands tested were substantially chloride free and suitable for use in **SSWs** in cleanrooms where corrosion of silicon wafers could not be tolerated.

18. Air Guns

Problem: Were two brands of nuclear ionized air guns equivalent in performance? What was a safe distance, in terms of the **field,** from these guns and from two brands of guns of the electrical type?

Solution: The approach was to use **special tests.** A 6 × 8 inch piece of FR-4 epoxy-glass circuit-board laminate was given an **apparent charge** of −5000 V by **triboelectric charging** and blown with the competitive nuclear air guns at 3 feet. After one second, the residual **apparent charge** was −4000 V in both cases. After 2 seconds, one was −1500 V, and the other was −1000 V. After 3 seconds, both were −500 V. The conclusion was that the brands were equivalent. Note, however, that certain polonium 210 ionizers were recalled in February 1988 because of an instance of escape of microspheres in which the radioactive isotope was encapsulated (Ref. 2-12). A brand of nuclear air gun using a different method of encapsulating the polonium-210 was still allowed at that time.

The **field** from nuclear ionizers is very low (Ref. 2-15 indicates charging of isolated **conductors** to less than 50 V), and a safe distance of 1 inch from the nozzle was established in **special tests** using **MOSFETs**. However, electrical ionized air guns caused **MOSFET** damage at 1 inch from the nozzle; the safe distance was established as 2 inches for one model and 3 inches for the other (Table 5 of Paper No. 3 in the Appendix).

19. Heat Guns

Problem: A heat gun was suspected of charging electrically isolated **conductors** so that **ESDS items** might be damaged by the **DI** mechanism.

Solution: The approach was to use **special tests.** The tip of the heat gun was held 4.7 inches from a vertical 12 × 12 × 1/32 inch aluminum sheet on a stand-off (plastic beaker). A vertical 6 × 8 inch section of aluminum window screen, also on a stand-off, was placed midway between the gun and plate. The gun was set on minimum air intake with the heater either on or off. The following data are for duplicate tests:

		VOLTAGE ON SHEET BY **FIELD METER**	
HEATER	SCREEN	AFTER 30 SECONDS	AFTER 60 SECONDS
Off	Ungrounded	0, 0	0, 0
On	Ungrounded	+550, +510	+650, +650
On	Grounded	+210, +230	+300, +300

Similar positive charging by a heat gun is reported in Ref. 2-15.

In another test, the gate lead of a Motorola 2N4351 **MOSFET** was connected to (1) a 2.5 × 4 inch copper "antenna" on FR-4 epoxy-glass circuit-board laminate, or (2) the same with an additional "antenna" (1 ft^2 aluminum sheet) attached, or (3) a 9-ft length of 12-AWG Teflon-insulated wire with a section of the insulation removed where heat-shrinkable tubing was to be installed. The substrate-case lead of the **MOSFET** was grounded. Results follow:

GATE LEAD CONNECTION	OPERATION	**MOSFET** DAMAGE
1	Hot air blown onto "antenna" for 1 min with tip of gun 5 inches from surface	0/5
2	Hot air blown onto aluminum plate in the same manner	5/5 (0S)
3	Section of heat-shrinkable tubing installed on bare conductor with tip of gun 3 inches from tubing, hot air blown for 1 minute	2/5 (0S)

Damage to the **MOSFETs** was arbitrarily defined as a shift in $V_{GS(TH)}$ of more than 0.03 V. The designation "2/5 (0S)," for example, means that 2 of 5 **MOSFET**s tested were damaged (voltage shifts of more than 0.03 V), but none was shorted (S = shorted). Note that the **MOSFET** we used is highly susceptible to ESD damage not only because of its low **HBM** threshold, about 100 V, but because it is voltage-sensitive rather than current-sensitive. Thus, a low-**capacitance** charged wire or circuit element might damage **MOSFET**s but be harmless to current-sensitive devices.

The data showed that significant voltages were produced on an electrically isolated aluminum plate when the gun blew hot air but not when it blew room-temperature air. Positive ions in the hot air were involved in the charging process as evidenced by about 50% less charging when a screen in the air stream was grounded than when it was ungrounded; presumably half the ions were caught (neutralized) by the screen and half passed through the openings. Apparently the heating element (high-resistance wire at red heat) was creating ions. The same positive charging by a heat gun was noted in Ref. 2–15, which states that "the charge generation occurs with the temperature rise of the heating elements." Aclar plastic was also charged by hot air from the gun in our tests, as was Teflon in Ref. 2–15. The damaging effect of the charges was proved in the **MOSFET** tests, the most realistic of which involved actual shrinkage of sleeving onto wire, which is the purpose of the heat gun. Conclusions were:

1. The subject heat gun was an ESD hazard for **MOSFET**s sensitive to 100 V.
2. Positive ions in the air stream were involved in charging the surface being heated.
3. Worst-case laboratory tests should be run with the specific **ESDS** devices installed in hardware on which the heat gun might be used. For example, the gun might prove safe for 1000-volt-sensitive devices. However, a surer solution to the problem would be to replace the gun by a different heating tool such as a Glo-Ring which depends on radiant heating rather than blowing hot air and is said to be "ideal for many applications including heat-shrinking."

20. Triboelectric Charging by Air

Problem: Pure gases such as air cannot cause **triboelectric charging** (Ref. 2–14), but might real air, with its dust content, cause charging as has been claimed (Refs. 2–4 and 5–3)?

Solution: The approach was a **special test.** A cellulose acetate butyrate tote box lid, a piece of **nonconductive** polyethylene bubble-wrap, and a piece of FR-4 circuit-board laminate were left in a circulating-air oven with an air flow of 26 feet/second at 200°F. The ambient relative humidity was 48% at 72°F. After 2.5 days in the oven, the three plastic pieces were removed and found to have less than 100 V **apparent charge,** which is harmless on **nonconductors** (Fig. 3 of Paper No. 3 in the Appendix). The conclusion was that blowing air at 200°F did not cause **triboelectric charging** in ovens of the kinds tested. However, note that a heat gun did produce ions, possibly emitted from the hot filament (Example 19, above). Also circuit-board laminates were charged to + 1500 V **apparent charge** by **triboelectric charging** when they slid on stainless-steel oven shelves (see the complete version of Paper No. 8 condensed in the Appendix), and some observers may have falsely attributed such charging to hot, blowing air. Electrically unbalanced air (with a preponderance of either positive or negative ions) is another possibility.

21. Charging of Electrically Isolated Conductors by Ionizers

Problem: What was the safe working distance for **ESDS items** from an AC ionizer or a pulsed DC ionizer? The latter had been observed to cause sparks to be thrown between electrically isolated **conductors,** e.g., metallic bases of microscopes on rubber feet, and grounded **conductors.**

Solution: The approach was to use **special tests.** When a **MOSFET** with a relatively large circuit line as an antenna was brought near either type of ionizer, damage occurred at 2 or 4 inches but not at 10 inches, so 10 inches was judged to be a safe working distance (Table 5 of Paper No. 3 in the Appendix). In another test, an electrically isolated aluminum plate (12 × 12 × 0.025 inch) was allowed to sit in the air stream for 1 minute before being touched by the gate lead of a **MOSFET,** with the substrate-case lead being held by a grounded operator. This time the AC ionizer caused damage at 8 inches (2/2 **MOSFETs** damaged) but not at 12 inches (0/5 damaged), so the safe working distance was revised slightly upward from 10 to 12 inches. However, the pulsed DC ionizer at 22 cycles/minute caused damage at 12 inches (3/3, 2 shorted), 18 inches (1/5, 1 shorted), and 24 inches (1/5, none shorted), but not at 36 inches (0/5). At 36 cycles-minute the damage was 3/3 (2 shorted) at 12 inches, 2/3 (1 shorted) at 18 inches, and 1/2 (none shorted) at 24 inches, so increasing the pulse rate from 22 to 36 cycles/minute made no noticeable improvement.

In conclusion, 1 foot was established as the safe working distance for AC ionizers (when balanced as was the one tested) and 3 feet for the pulsed DC ionizer. But 3 feet was too far to be practical, so the pulsed DC ionizer was put out of service; all units were collected and discarded.

GIDEP Alert E9-A-86-011 stated that pulsed DC ionizers did not protect **ESDS** devices, but later testing is said to have shown no ion imbalance at 6 inches with the ionizer operating at 5 Hz (cycles/second) and an imbalance of only 60 V at 2 Hz; no **MOSFETs** (Texas Instruments 3N163) were damaged (Ref. 5-4). However, **MOSFETs** were damaged at 0.4 Hz (22 cycles/minute) as stated in the GIDEP Alert and found in our tests, above. It would seem that the frequency must be about 5 Hz, minimum. We advise the potential buyer of pulsed DC equipment to satisfy himself, preferably with his own tests, that isolated **conductors** won't be charged to cause damage by **DI.**

22. Old AC Ionizers

Problem: Were old bench-model AC ionized air blowers still well enough balanced electrically to effectively neutralize charges and to be safe at 12 inches in terms of charging isolated **conductors**?

Solution: The approach was to use **special tests** with a Model 200 Charged Plate Monitor from Ion Systems, Inc. For one ionizer, the time to discharge the plate from −1000 V to −100 V at 2 feet was found to be 4 seconds, and from +1000 V to +100 V the time was 5 seconds. These data indicate that positive ions were more numerous than negative ions in the air stream, and in fact the isolated plate was charged to +80 V at 1 foot and +59 V at 2 feet. The conclusion was that this ionizer, whose points were rounded and dirty, was still effective but was on the borderline of damaging 100-volt-sensitive **ESDS items** at 1 foot via isolated **conductors.** A new ionizer of the same model charged the plate only a few volts at one foot.

A second model of ionizer, which was several years old but had been given light usage, charged the plate only about +20 V at 1 foot and was judged safe at that distance for **ESDS items** sensitive to 100 V.

23. Fanless Ionizer

Problem: Was a novel fanless, nonpulsed, two-point DC ionizer sufficiently effective to be substituted for conventional AC ionizers with fans?

Solution: The approach was a **special test. Triboelectric charging** of **nonconductive** polyethylene bubble-wrap with polyester fabric gave −3300 V **apparent charge** on the bubble-wrap, which was placed 2 feet in front of either the DC ionizer or a conventional AC ionizer for either 30 or 60 seconds. At 30 seconds, the residual charge was −3200 V with no ionizer, −2600 V with the DC ionizer, and −1200 V with the AC ionizer. At 60 seconds, the residual charge was −3000 V with no ionizer, −1700 V with the DC ionizer, and −700 V with the AC ionizer. The conclusion was that the novel DC ionizer was less than half as effective as the conventional AC ionizer and was not an acceptable substitute.

24. Grit-Blasting of Circuit-Board Laminate

Problem: A Micro Blaster was found to create +700 V **apparent charge** on FR-4 epoxy-glass circuit-board laminate being sprayed with sodium bicarbonate powder to remove conformal coating. How could this unacceptably high charge be reduced?

Solution: The approach was a **special test.** A "piggyback" corona-discharge ionizer reduced the **apparent charge** to +60 V (see the "Assembly Operations" section of Paper No. 3 in the Appendix), which is harmless on **nonconductors** because the **field** is far too weak to damage **ESDS items** by the **FIM** (see under "Field Strength versus Distance" in the same paper).

25. Grit-Blasting of Heat Sink on Module

Problem: The nickel plating on the aluminum heat sink on a module was peeling, so the plating was removed by grit-blasting with alumina powder. The heat sink was grounded, and the rest of the module was protected from the grit by metal shields taped in place. The result was ESD damage to all 83 CMOS devices on the board, so that the $150,000 module had to be scrapped. What ESD precautions should have been taken in the grit-blasting process?

Solution: The approach was **analysis;** recommendations were made on the basis of general knowledge without running tests. The recommendations were: (1) ground the shields and the ground planes of the board as well as the heat sinks, (2) use an air ionizer in conjunction with the Micro Blaster (as in Example 24, above), (3) use **conductive** tape, e.g., aluminum foil, to hold the shields, and (4) monitor the **apparent charge** with a **field meter** during the grit-blasting process.

With the recommended precautions, no device was damaged. This is a dramatic case of the need for ESD control. The loss of $150,000 (1986 money) was certainly an attention-getter for management. And note that such overt failure suggests the possibility of undetected damage leading to **latent failure** when ESD precautions are lax.

26. Grounding versus DI Damage

Problem: Was **grounding** necessary for a **MOSFET** on a model circuit board to be damaged by the **DI** mechanism? The answer was wanted for general information.

Solution: The approach was a **special test** using the **MOSFET** board described under "MOSFET Damage Tests" in Paper No. 3 in the Appendix. A piece of aluminum foil (9 × 11 inches) was charged to +4000 V from a power supply and touched to the lead of the **MOSFET** board which was supported by an **insulator.** The **MOSFET** on the board was damaged. Similar tests gave the same result. The conclusion was that voltage-sensitive devices, which need little current flow to be damaged, are susceptible to **DI** with none of the pins being grounded. Note that a similar result was found for **FFB** damage (see the complete version of Paper No. 8 condensed in the Appendix); here, again, no grounding was needed for **MOSFET** damage, but the damaging charge came from a **field** rather than from **DI.**

27. Ultraviolet Light

Problem: Ultraviolet (UV) light is known to cause the development of a positive charge on surfaces by the photoelectric effect. Is inspection of **ESDS items** under UV light a safe process?

Solution: The approach was a **special test.** A **MOSFET** board (Paper No. 3 in the Appendix), coated with Humiseal 1B-31 acrylic conformal coating, was exposed to UV light from a Blak-Ray B-100A lamp (Ultraviolet Products, Inc.) at 12 inches for 5 minutes. The **MOSFET** was undamaged. A total of three tests gave the same result (0/3 damage). The conclusion was that inspection of the coating, which took less than 5 minutes, was not an ESD hazard (assuming that standard precautions, e.g., **grounding** of the operator, were taken during this inspection).

28. Desiccator Cabinet

Problem: Could **static charges** develop on **nonconductors** lying inside a desiccator cabinet under dry, slowly flowing nitrogen? The answer was desired for general information.

Solution: The approach was a **special test.** Three 6 × 10 inch specimens were placed in the cabinet: FR-4 circuit-board laminate, polyethylene film, and polystyrene foam sheet. A nitrogen flow of 10 standard cubic feet/hour was maintained for 24 hours. At the end of this time the specimens were removed, taking care not to slide them, and measured with a **field meter.** The **apparent charge** was less than 100 V in all cases.

The conclusion was that **static charges** did not develop on **nonconductors** sitting in the cabinet for a day. On the other hand, **nonconductors** could be charged by sliding on the grounded steel shelves of the cabinet as observed for oven shelves and conveyor belts (see Example 20 above).

29. Chain Link Fence as Shield

Problem: A chain link fence separated **SSW**s from an aisle where people in ordinary clothing were walking. How far from the fence should the **SSZ**s begin?

Solution: The approach was a **special test.** A piece of Aclar film was stroked with polyester fabric to give an **apparent charge** of −4000 V. When the charged Aclar was held against the fence (wire diameter 0.114 inch, 12 openings/3 feet measured horizontally), the **field meter** on the other side of the fence read −170 V. Thus the attenuation was 4000/170 = 24 times. In another test, the **apparent charge** was reduced from −10,000 V to −500 V by the fence, so the attenuation was 10,000/500 = 20 times. If the voltage were 10,000 V on passers-by, the safe distance from the fence by the **CD Rule** in inches would be $(500)^{0.5}/1.8 = 12$. If the voltage were 20,000 V on passers-by, which is considered a worst-case maximum, the safe distance in inches would be $(1000)^{0.5}/1.8 = 18$. In conclusion, **SSZs** were maintained at 18 inches, minimum, from the fence.

30. Packaging of Electronic Box with Connector

Problem: A metal-cased electronic box was a complete **Faraday cage** except for a connector with deeply set pins. Could the connector be left uncapped and the box shipped in a see-through metallized plastic bag so that the serial numbers on the box could be read without unpacking?

Solution: The approach was a **special test.** In a variation of the standard **discharge test** in paragraph 4.3.3 of Model Specification 2 in Chapter 6, a connector pin was connected to the gate lead of a **MOSFET** with the substrate-case lead being connected to the rim (backshell) of the connector. The **MOSFET** and the two wires on it were bagged in a foil laminate bag to protect them from a **field** during the test, and the metallized plastic was pulled tight across the connector face and touched with the probe at 10,000 V under the conditions of the low-resistance **discharge test** (paragraph 4.3.3(2) of the specification). This was a worst-case condition representing the finger of a charged person touching the wrapping. Five **MOSFETs** in a row were undamaged ($^0/_5$ damage), so the conclusion was that the metallized plastic bag was satisfactory for shipping (but it might not have been if the pins had been less deeply set in the connector housing). The particular combination of shipped item (electronic box) and packaging was an **SSP** as defined in Chapter 2; though a **special test** was done, it was construed as equivalent to the standard **discharge test** (in Model Specification 2) on which it was based.

31. Shipping Test

Problem: Would **MOSFETs** sensitive to 100 V be damaged by in-plant shipping in a 3-mil plain **nonconductive** polyethylene bag? The answer was desired for general information.

Solution: The approach was a **special test** using a variation of the **MOSFET** board (Paper No. 3 in the Appendix), with five **MOSFETs**, each with its own separate antennas (about 1.7 in.^{2} each). When a person held the bag with his fingers over the antennas, walked across a carpet, and handed the package to another person who seized it above the opposite antennas, a **MOSFET** was damaged by the charge on the fingers acting through the polyethylene. In a control test, **MOSFETs** were undamaged, as expected, using a foil laminate bag (a **Faraday cage**).

Next, the **MOSFET** board in the plain polyethylene bag was placed in a paper envelope and mailed 5 times across a corporate facility, involing sorting in a mailroom. None of the **MOSFETs** was damaged (0/5 damage).

The conclusion is that ESD damage is not inevitable even with very sensitive devices in a bag with minimal **shielding** capability, but even the remote possibility of damage of high-reliability parts is reason enough to use an **SSP.**

32. Plated Wire Memory Shipping Container

Problem: The subject container had a heavy aluminum wall which was anodized so that its surface was **nonconductive.** Thus the container was a **Faraday cage,** assuming proof of electrical continuity, but was not an **SSP** by company definitions requiring a **conductive** or **antistatic** inner surface. The container was used for shipment. Could the plated wire memory, which was **ESDS,** have been damaged?

Solution: The approach was **analysis.** Electrical continuity of the container, as required for a **Faraday cage,** had been verified by probing. Since the memory was rigidly mounted inside the container, no **triboelectric charging** could have occurred during shipping, and the conclusion was that the package violated the requirements of a company specification for an **SSP** but was satisfactory in practice. In fact, the heavy aluminum wall made the container virtually impregnable to **fields** and **discharges.** The plated wire memory was judged to have been ESD-safe.

By paragraph 3.4.2 of Model Specification 2 in Chapter 6, this container would have been allowed as an **SSP** because the packaged item was securely mounted to prevent **triboelectric charging** by sliding.

33. Various Packaging Configurations

Problem: Various packages proposed by suppliers were not **Faraday cages.** Might some of these give sufficient protection against **fields** and **discharges** by reason of configuration such as wide spacing between the **ESDS item** and the outside of the package?

Solution: The approach was to develop a **standard test,** namely the **discharge test** of paragraph 4.3.3 of Model Specification 2 in Chapter 6. Thus the definition of an **SSP** was broadened beyond the **Faraday cage** defined in paragraph 3.2.2 of Model Specification 2 in Chapter 6. Cardboard boxes lined with **antistatic** polyethylene foam were able to pass the **discharge test** without a **conductive** layer for **shielding.** The purpose of broadening the definition of an **SSP** was to allow suppliers latitude in their packaging methods, but a true **Faraday-cage** package, i.e., one with buried foil or screen, is preferred as explained in Chapter 3.

34. Silicone-Rubber Thermal Barrier on Probe

Problem: In the evaluation of complex hybrid circuits for dynamic parameters under temperature variations, a probe was encased in silicone rubber as a thermal barrier. Unexplainable malfunctions and intermittent failures were ascribed to the **field** from **apparent charges** of commonly −5000 V and sometimes as high as −15,000 V. How could **triboelectric charging** of the silicone rubber (or other probe-encasing rubber or plastic material) be controlled?

Solution: The approach was a **special test.** Since no high-temperature **antistatic** material with all the required properties for encasing the probe was known, the silicone rubber remained in use, but **ionization** was applied; the air flow from an AC ionizing blower 3 feet away was directed on the probe. Now the **apparent charge** was limited to at most −300 V, which is harmless on **nonconductors** (see Fig. 3 of Paper No. 3 in the Appendix).

35. Automated Process

Problem: How could the ESD-safety of a complex automated process, with various hazards involving all the ESD damage mechanisms **(DI, FIM, CDM, FFB)**, be assured with high confidence?

Solution: The approach was a combination of **analysis** and **special tests.** Paper No. 8 in the Appendix outlines how the problem was solved by the **coupon** approach, which is uniquely applicable to automation and gives very high confidence in ESD safety because of the absence of unpredictable operator error. Mishandling cannot occur unless the process goes awry, as would be detected in periodic **coupon** checks.

Note that the **coupon** procedure has been made part of Model Specification 1 (paragraph 3.9) in Chapter 6. This is another example of the way in which specifications grow by absorbing the solutions found by troubleshooting. In terms of this book, an expanding Chapter 5 continues to feed into Chapter 6.

Chapter 6

Model Specifications

Introduction

These model specifications are not claimed to be "perfect" or all-inclusive. Their purpose is to illustrate and codify the elements of our approach, e.g., the **SSZ** concept and the **CD Rule**, in specification format. Here the threads of our preceding A-to-Z discussions are pulled together in a system of required procedures. Note how the results of special studies (troubleshooting) have been incorporated as pointed out at the end of the last chapter.

Specifications may be written as a list of general requirements without "how-to" instructions or detailed examples, but the problem with this method is that the reader is puzzled by sterile, though correct, generalizations. Alternatively, specific materials, procedures, etc., can be given, but the problem here is lack of flexibility for meeting general requirements in new, creative ways. Our specifications are compromises; we give a general requirement and then illustrate it with specifics. Thus the intent is stated along with methods—but not the only methods—to achieve it.

Model Specification 1 concerns the **SSW** and gives rules for maintaining an **SSZ** within it using the various "tools" of Chapter 2. Similarly, Model Specification 2 utilizes these tools for the **SSP.** You should be able to adapt these specifications to your own use by appropriately modifying and expanding them. Use them freely; see the Copyright page.

Model Specification 1. Handling and Assembly of Unpackaged ESDS Items

1. Scope

This specification establishes requirements for in-plant handling aspects of an ESD-control program as required by MIL-STD-1686A and contains detailed requirements which are called for but not given in that document.

The purpose of the ESD-control program is to protect electrical and electronic items (including components, assemblies, and units as defined in 5.1) that are susceptible to damage or degradation by ESD. The ESD-control program does not apply to explosives or other hazardous materials.

2. Applicable Documents

The following documents, of the latest issue in effect except as otherwise indicated, form a part of this specification to the extent specified herein. In the event of conflict between documents referenced herein and the contents of this specification, the contents of this specification shall govern.

2.1. Government documents.

SPECIFICATIONS	
Military	
MIL-M-38510	Microcircuits, General Specification for
STANDARDS	
Military	
MIL-STD-129	Marking for Shipment and Storage
MIL-STD-1686A	Electrostatic Discharge Control Program for Protection of Electrical and Electronic Parts, Assemblies, and Equipment (Excluding Electrically Initiated Explosive Devices)
DoD-STD-2000-1B*	Soldering Technology, High Quality/High Reliability
OTHER PUBLICATIONS	
DoD-HDBK-263	Electrostatic Discharge Control Handbook for Protection of Electrical and Electronic Parts, Assemblies, and Equipment (Excluding Electrically Initiated Explosive Devices)
NAVSEA SE 003-AA-TRN-010	Electrostatic Discharge Training Manual

*Note: Superseded by MIL-STD-2000 on January 16, 1989. We reference DOD-STD-2000 because many older Government programs are on contract for it.

2.2. Nongovernment documents.

SPECIFICATIONS

Model Specification 2	Packaging of **ESDS Items** for Shipment, Storage, and In-Plant Transfer

OTHER PUBLICATIONS*

Reference Document 1	General Design, Construction, and Maintenance Guidelines for ESD-Protected Areas
Reference Document 2	Certification of Static-Safe Workstations **(SSWs)**
Reference Document 3	Procurement Requirements for Continuous Wrist-Strap Monitor Unit
Reference Document 4	Approved ESD-Control Materials and Equipment
Reference Document 5	Criteria for Selection of Approved ESD-Control Materials and Equipment
Reference Document 6	Certification of Personnel for ESD Control

3. Requirements

3.1. Order of precedence. In the event of conflict between the requirements of this specification, a program contract, and drawings calling out this specification, the requirements of the contract and the drawings shall take precedence in that order.

3.2. Classification of ESDS items. ESDS items shall be classified in accordance with MIL-STD-1686A. Items in Classes 1–3 shall be considered **ESDS** for the purposes of this specification, and all shall be protected in full accordance with this specification.

3.3. Design protection. The need for design protection shall be determined in accordance with DoD-HDBK-263 and NAVSEA SE 003-AA-TRN-010.

---

*Reference Documents 1–6 are made independent of the specification so that they can be revised separately. This is especially important for Reference Document 4, which is a "living document" that is always growing and changing.

3.3.1. *Protection of components and assemblies.* Class 1 components shall not be used where Class 2 or less sensitive components are available that will meet performance requirements. Where Class 1 components must be used, protective circuitry shall be incorporated at the lowest practical level of assembly to limit the sensitivity of that assembly to Class 2 or greater. Where protective circuity cannot be used, reasons shall be presented to the acquiring activity.

3.3.2. *Protection of external unit enclosure terminals.* Unit terminals, including test points, located on the surface of the unit enclosure shall contain protective circuitry making the unit sensitive to no less than 4000 V (Class 3). When protective circuitry cannot be used to provide this protection, reasons shall be presented to the acquiring activity.

3.4. Duration of protection. **ESDS items** shall be protected in accordance with 3.5 from the time of receipt through final unit assembly, or until processing eliminates the possibility of ESD damage as defined by the drawing. Items whose electrical parameters are irrelevant so that ESD damage is of no concern shall be exempt from this requirement and from all requirements of this specification; examples are display models and scrap items used for physical as opposed to electrical testing.

3.5. Static-safe zone (SSZ). While being protected (3.4), **ESDS items** shall never leave an **SSZ,** which is defined as a volume in space at every point of which the **Basic Rule** (3.5.1) is followed.

3.5.1. **Basic Rule.** An **ESDS item** shall never, even for a nanosecond, be exposed to an **E field** in violation of the **CD Rule** (3.6.11) or have its sensitive leads or terminals touched to, or receive a **discharge** from, any surface at more than 50 V.

3.5.2. *Location of SSZ.* The **SSZ** may be (1) inside a static-safe package **(SSP)** as defined by Model Specification 2, (2) within a static-safe workstation **(SSW)** as defined in 3.6.1, or (3) around the **ESDS item** while this item is being relocated from one **SSW** to another under the continuous supervision of ESD-certified personnel (3.6.13). An **ESDS item** shall be removed from an **SSP** only within the **SSZ** of an **SSW.**

3.6. Static-safe workstation (SSW)

3.6.1. *Definition of* **SSW.** An **SSW** is a volume in space which is designated and equipped to maintain an **SSZ** (3.5).

3.6.2. *General design, construction, and maintenance.* General design, construction, and maintenance of **SSW**s and ESD-protected areas in general shall be in accordance with Reference Document 1.

3.6.3. *Identification and numbering of* **SSWs.** All **SSW**s shall be identified by signs and markings, suitably strips of yellow tape at the boundaries, and shall be numbered for the purpose of keeping records, e.g., of certification (3.6.4).

3.6.4. *Certification.* **SSW**s shall be certified annually in accordance with Reference Document 2.

3.6.5. **Grounding**

3.6.5.1. **Grounding** *of* **conductors.** All **conductors** shall be grounded so that their potential versus earth ground never exceeds 50 V as required by the **Basic Rule** (3.5.1).

3.6.5.2. **Grounding** *of work surfaces and personnel using static ground.* A static ground separate from commercial power ground shall be required for reasons of safety. For example, the commercial power ground might be floating (disconnected) and powered by a hot lead so that operators are endangered.

3.6.5.2.1. *Static-ground terminations.* Static-ground terminations shall be one of the following:

a. Building structural steel.
b. Clamp to a metallic cold water pipe.
c. Wet ground-well with copper-clad steel stake.

(Test and process equipment chassis shall be connected to commercial power ground.)

3.6.5.2.2. *Voltage and resistance differential from commercial power ground.* Static ground, measured at the distribution wire (3.6.5.2.3 and Fig. 1), shall not differ from commercial power ground by more than 5 V 9dc or rms) or by more than 25 ohms.

3.6.5.2.3. *Static-ground wire system.* The grounding lug of the work surface shall be connected by a lead wire with a resistor (3.6.5.2.4) to a distribution wire terminating in ground (3.6.5.2.1). The lead wire shall be a minimum of 18 AWG, and the distribution wire shall be a minimum of 12 AWG. (AWG = American Wire Gauge.) Fig. 1 shows a typical **grounding** system.

3.6.5.2.4. *Resistor.* There shall be no resistor between the termination and the distribution wire, but there shall be a 1-megohm resistor between the distribution wire and the **grounding** lug of the work surface.

3.6.5.3. **Grounding** *of personnel.*

3.6.5.3.1. *Personnel protection.* Personnel shall be grounded by static ground (3.6.5.2) and shall always be protected by sufficient resistance to limit current to a maximum of 1.0 mA when they are electrically connected to ground, e.g., by a wrist strap. (Current in mA = 1000 × voltage/resistance in ohms.)

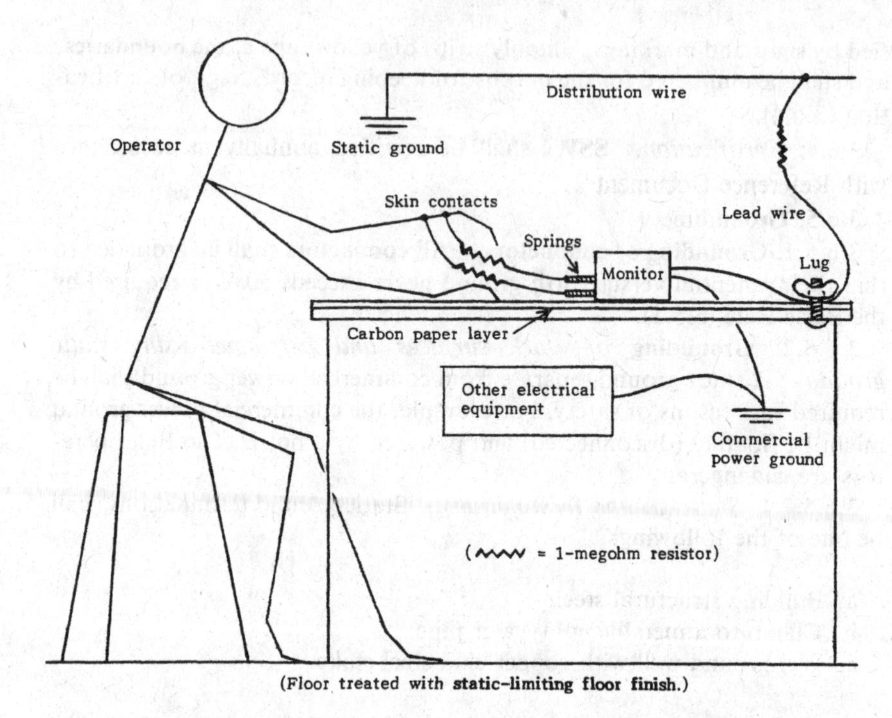

Fig. 1. Diagram of **grounding** of a typical **SSW** with workbench.

3.6.5.3.2. *Allowable resistance to ground* **(ARTG).** The **ARTG,** as measured at 0.1–40 V dc between static ground and a metallic probe 1–3 inches in diameter held firmly in the operator's hand, shall be a maximum of 10 megohms and a minimum of 1 megohm.

3.6.5.3.3. *Wrist-strap monitor.* Every wrist strap shall be monitored by a **continuous wrist-strap monitor** of the resistive type procured in accordance with Reference Document 3. One ground connection of the monitor shall be connected to static ground, suitably by means of the grounding lug of the work surface, and the other ground connection of the monitor shall be connected to commercial power ground. See Fig. 1. (*Note:* Some users have found continuous-monitor features such as the audible alarm to be annoying. As we say in Paper No. 10 in the Appendix, don't buy these monitors, or any other major ESD-control equipment, without a trial in your own assembly area.)

3.6.5.3.4. *Ground fault circuit interruptors (GFCIs).* GFCIs shall be installed when voltages above 225 V are being used, so that personnel will be protected even if electrically overstressed resistors should be carbonized so that they allow a current of more than 1 mA.

3.6.5.3.5. **Conductive** *floor mats.* **Conductive** floor mats or other methods of **grounding** personnel other than monitored wrist straps (3.6.5.3.3) shall be used only when necessary. An example is the **grounding** of mobile personnel when wrist straps even with long cords or connected to a sliding trolley overhead are impractical. Floor mats shall be connected to the static-ground distribution wire (3.6.5.2.3) through a 1-megohm resistor, or they may be connected directly to the **grounding** lug of the work surface. **Conductive** shoes or shoe straps with a minimum resistance of 1 megohm shall be used in conjunction with **conductive** floor mats.

3.6.6. *Materials and equipment.*

3.6.6.1. *Materials.* ESD-protective materials used in accordance with this specification shall have a **surface resistivity** of less than 10^{12} ohms/square when measured at 70°F and a maximum of 50% relative humidity with a Voyager SRM-110 **Surface Resistivity** Meter or equivalent. Materials shall be selected from Reference Document 4, which is a listing prepared using the criteria of Reference Document 5. The materials are classified as follows:

Conductive (surface resistivity less than 10^5 ohms/square).

Static-dissipative (surface resistivity at least 10^5 but less than 10^9 ohms/square).

Antistatic (surface resistivity at least 10^9 but less than 10^{12} ohms/square). Unglazed paper is antistatic at normal relative humidity (above about 25% at 70°F) and is acceptable for work instructions and other documents in the **SSW**. Unglazed cardboard is also **antistatic** and may be used for tote boxes, etc. (*Note:* EIA 541 combines **antistatic** and **static-dissipative** as "dissipative," which is sensible as we mention elsewhere, but we follow DoD-HDBK-263 practice here.)

3.6.6.2. *Equipment.* Equipment used in accordance with this specification shall be selected from Reference Document 4, which is a listing prepared using the criteria of Reference Document 5.

3.6.7. *Relative humidity.* If the relative humidity is controlled in order to reduce **triboelectric charging,** the absolute humidity shall be equivalent to a relative humidity of 30%, minimum, at 70°F. (*Note:* 40% would be even better. The maximum must be set with consideration of possible adverse effects such as corrosion, delamination of circuit-board material, or personnel discomfort.)

3.6.8. *Ionization.* Air ionizers shall be used only when absolutely necessary, e.g., in grit-blasting or spray-coating operations. Equipment shall be selected from Reference Document 4. Ionized air blowers shall be electrical, but nozzles (guns) may be either electrical or nuclear. (*Note:* This is our preference based on cost of use, because the nuclear-type blowers are leased and have an annual cost that accumulates excessively, in our opinion.) The blowers shall have autobalancing, point-**discharge** indication, and need-for-cleaning indication features. All **ionization** equipment shall be tested and certified annually. **ESDS items** shall never be nearer than 12 inches to the points of a bench-model ionized air blower or 3 inches from a nozzle; this prohibition protects the items from slight electrical imbalance of the ionizers. (*Note:* Even nuclear (polonium-210) blowers have been accused of creating space charges and charging isolated **conductors** to a little under 50 V (Ref. 2–15).)

3.6.9. **Static-limiting floor finish.** An approved **static-limiting floor finish** (Reference Document 4) shall be applied to all tile and concrete floors (over a sealer for the concrete) and maintained so that appearance is attractive (gloss = 25, minimum; see Paper No. 6 in the Appendix), slip resistance is adequate (0.20 Topaka, minimum; see Paper No. 6 in the Appendix), and the **surface resistivity** never exceeds 10^{11} ohms/square as measured by a Voyager SRM-110 **Surface Resistivity** Meter or equivalent. Carpeted floors shall be used only when absolutely necessary, and the carpet shall be treated with an approved antistat; the limitation of charges on people shall be measured by the **walk test** of Paper No. 6 in the Appendix or by measurements with a Voyager Personnel Voltage Tester PVT-300 or equivalent.

3.6.10. *Necessary* **nonconductors.** **Nonconductors** in the **SSW** shall be limited to those which are not easily or cost-effectively replaced. Examples are:

1. Painted, grounded metals
2. Plastic-coated, grounded metals with plastic less than 20 mils thick and metal more than 60 mils thick
3. Latex gloves (see 3.6.12.11)
4. Floors
5. Chairs
6. Walls
7. Structural members within the workstation
8. Personal apparel including I.D. badges
9. Insulated (**nonconductive**) tool handles
10. Dispensing syringes and bottles

As necessary, these shall be treated with **topical antistat** to make them **antistatic** so that their **field** will not violate the **CD Rule** (3.6.11). If corrosion is feared, a chloride-free **topical antistat** may be selected. In the above list, Nos. 1 and 2 will not need antistat treatment beacause of **voltage suppression,** No. 3 will be used without antistat treatment but subject to the conditions of 3.6.12.11, Nos. 4–8 probably will not need antistat treatment (for badges, see Example 4 in Chapter 5), and Nos. 9 and 10 shall be treated with **topical antistat** and so labeled. The treatment shall be renewed every three months (*Note:* or more often, as indicated by experience), and its effectiveness shall be monitored by **field-meter** checks.

3.6.11. **CD Rule.** The **CD Rule** shall never be violated, the requirement being that **ESDS items** must be kept at least a safe distance d from a surface with **apparent charge** V in volts according to the equation: $d = V^{0.5}/1.8$. The **apparent charge** is the charge in volts of a surface or object as read by a handheld **field meter** according to the meter manufacturer's instructions; this is an uncorrected reading and represents the **field** strength at the point in space where the meter is held. (*Note:* See the discussion of the **CD Rule** in Chapter 2.) For convenience, some solutions of the equation are:

V, VOLTS	d, INCHES	V, VOLTS	d, INCHES
300	10	6000	43
500	12	8000	50
1000	18	10,000	56
2000	25	15,000	68
3000	30	20,000	79
4000	35	30,000	96

3.6.12. *Operator disciplines.*

3.6.12.1. *Maintenance of* **SSZ.** The operator shall maintain an **SSZ** and shall always know its boundaries so that **ESDS items** are kept within it. In general, an **SSZ** will end at an operator's wrists, at least 12 inches from the points of an ionized air blower or 3 inches from a nozzle (3.6.8), and at distance d from any charged surface according to the **CD Rule** (3.6.11).

3.6.12.2. *Parts handling.* The operator shall not unnecessarily touch **ESDS** leads, terminals, or contacts even though he or she is grounded. Parts shall be handled by their cases as size permits. An **ESDS item** shall never be held beyond the front edge of the work surface because of possible **fields** from the chair or operator's lower garments.

3.6.12.3. *Touching of surfaces.* The operator shall not allow **ESDS items** to touch surfaces or objects other than the approved work surface, approved materials and equipment, or the assembly to which the item is being attached. For example, clothing shall not be touched.

3.6.12.4. **Triboelectric charging** *by body movement.* The operator shall not fidget, squirm, or shuffle his or her feet unnecessarily. The object of this discipline is to minimize **triboelectric charging** of the operator's body or apparel.

3.6.12.5. *Voltage surges by capacitance drop.* The operator shall not touch **ESDS** leads, terminals, or contacts while arising from a chair or lifting his or her feet. This prohibition protects **ESDS items** from surges in V when C drops in the equation $Q = CV$.

3.6.12.6. *Hair.* Hair shall be tied back as necessary so that it is within 2 inches from the skin in order to limit the **apparent charge** by **voltage suppression.** (*Note:* See Example 2 in Chapter 5. This requirement on hair is an example of the incorporation of the results of troubleshooting studies into the specification.)

3.6.12.7. *Shorting clips and protective containers.* The operator shall not remove shorting clips or protective containers until ready to use an **ESDS** device.

3.6.12.8. *Exclusion of* **nonconductors.** Static-generating **nonconductors,** e.g., ordinary, untreated tote boxes, plastic envelopes or notebooks (for work instructions, etc.), glazed paper, or ordinary plastic packaging material, shall be excluded from the **SSW.** Only necessary **nonconductors** (3.6.10) shall be allowed.

3.6.12.9. *Cleanliness.* A clean and orderly **SSW** shall be maintained. Lunches, coffee cups, candy wrappers, etc., shall be excluded; many of these materials are unnecessary **nonconductors.**

3.6.12.10. *Wearing apparel.* Operators shall wear (1) close-fitting blouses or shirts with short sleeves (above the elbow), (2) close-fitting blouses or shirts with sleeves rolled up above the elbow, or (3) ESD-protective smocks. (*Note:* In assembly areas, these smocks are an attractive "uniform" that raises ESD-awareness and may impress visitors, but periodic cleaning and replacement of the smocks may prove relatively expensive.)

3.6.12.11. *Gloves and finger cots.* Cotton or other **antistatic** gloves and finger cots shall be worn, with the exception that **nonconductive** gloves or cots, e.g., latex gloves for cleanroom work where **antistatic** additives might rub off to cause contamination, may be used when they are tightly fitting and are judged to be necessary **nonconductors** (3.6.10(3)). Of course, **nonconductive** gloves shall not be put on or removed at the **SSW** in such a way that their **field,** uncontrolled by **voltage suppression** as it is when they are being worn, violates the **CD Rule** (3.6.11). In general, **nonconductive** latex gloves shall be put on or removed at least 5 feet from the boundary of the **SSW.** (*Note:* For precautions on using these gloves, see Example 15 in Chapter 5.)

3.6.12.12. *Working near chain link fences.* The **SSZ** (3.5) shall be at least 18 inches from a chain link fence bounding the **SSW** when people in ordinary clothes may pass near the other side of the fence. (*Note:* This paragraph would not be included in most specifications. It's an example of a special provision based on an unusual situation, viz., the presence of fences in assembly areas; see Example 29 in Chapter 5.)

3.6.12.13. *Transient personnel.* Operators shall ensure that transient personnel, e.g., supervisors or visitors, do not enter **SSWs** unless such personnel are grounded and closely supervised by the operator.

3.6.13. *Certification and training of personnel.* Operators and other personnel who specify, acquire, design, assemble, inspect, test, package, repair, rework, install, or maintain **ESDS items** shall be certified in accordance with Reference Document 6 by completing a training course in ESD control (at least 8 hours instruction and one test), then recertified annually (at least 2 hours brush-up instruction and one test). Records of certification and recertification shall be maintained. (*Note:* Operator training should correspond to the specification; for example, if this model specification were used, our book with appropriate explanations and enlargements would serve as the text or basis of course material. The need for consistency of specification and training material seems obvious, but in practice there may be contradictions because the ESD-control program was developed piecemeal and the training course and the evolving specification went in different directions. Another point we wish to emphasize is that operators should not be kept ignorant by a management jealously guarding facts because "knowledge is power." Rather, in regard to ESD control the operator should be a "senior employee" able to act independently and intelligently like an airplane pilot or police officer.)

3.6.14. *Testing.*

3.6.14.1. *Dielectric strength or insulation resistance tests.* Equipment containing **ESDS items** shall not be subjected to dielectric strength or insulation resistance tests.

3.6.14.2. *Applied voltage.* **ESDS items** shall not be subjected to applied voltage, e.g., from an ohmmeter, at an energy level that could create a damaging current.

3.6.14.3. *Removal from test sockets.* **ESDS items** shall not be removed from test sockets with power applied.

3.6.14.4. *Power to connectors.* All power shall be turned off to the connectors in a system before printed circuit boards or other assemblies containing **ESDS items** are inserted into or removed from the connectors.

3.6.14.5. *Proper polarity of voltage.* Test equipment setups shall be arranged for proper polarity of voltage before conducting parametric or functional testing.

3.6.14.6. *Supply voltages and signal voltages.* Direct-current (dc) supply voltages shall always be applied before signal voltages except for memory subassemblies. Signal voltages shall always be removed before dc supply voltages unless the function removes signal voltages simultaneously with dc supply voltages.

3.6.15. *Installation site.* The following minimum installation procedures shall be enforced:

a. Protective coverings shall be maintained intact when **ESDS** units are to be stored prior to installation.
b. Protective caps, shunts, etc., on unit receptacles shall not be removed during unit installation until the completion of required interconnecting cable and connector assemblies.
c. Connector and receptacle surfaces shall be momentarily contacted to ground in order to discharge any electrostatic potentials on the cable or installer's body just prior to engaging a cable connector with a mating receptacle and prior to removing protective covering.

3.6.16. *Caution marking.*

3.6.16.1. *Labeling.* **ESDS items** or their containers shall be identified with an ESD caution label for all in-plant operations. The label may be on ordinary paper or latex-impregnated paper for cleanroom use and shall be backed with pressure-sensitive adhesive. The label shall include the MIL-STD-129 symbol or the Electronics Industries Association (EIA) triangle and shall be readily visible to personnel when marked assemblies are mounted on the next higher assembly.

3.6.16.2. *Caution tag.* When labels are not practical, an ESD caution tag shall be used. For cleanroom use, the tag shall be sealed in an envelope of transparent **antistatic** polyethylene or, better, nylon. Specific instructions may be recorded on the reverse side of the tag. A sample tag design follows.

3.7. Audits and reviews. Audits and reviews of the ESD-control program shall be conducted in accordance with contractual requirements. (*Note:* A thorough audit can be a disturbing revelation. The **ARTG** may be found to be exceeded in a majority of cases if **continuous wrist-strap moni-**

tors, as required in this specification, are not used, and many **SSWs** may be found to have faulty **grounding.** An adverse audit could result in much assembled hardware being considered suspect and perhaps rejected.)

3.8. Packaging for delivery. Packaging for delivery shall be in accordance with Model Specification 2.

3.9. Automated processes. Automated processes shall be certified ESD-safe, and periodically recertified, by means of a **coupon.** (*Note:* The particular coupon design and its use must be detailed in your own specification. See Paper No. 8 in the Appendix for a full discussion of the **coupon** technique.)

3.10. Disposition of mishandled hardware. As soon as possible after the infraction, the ten data items listed in Chapter 8 shall be recorded. Disposition of the hardware then shall be decided by Quality Assurance in conjunction with Engineering.

4. Quality Assurance Provisions

4.1. Responsibility for quality assurance. The Quality Assurance organization shall be responsible for compliance with the requirements specified.

4.2. Certification. **SSWs**, **ionization** equipment, and personnel shall be certified in accordance with 3.6.4, 3.6.8, and 3.6.13, respectively.

4.3. General inspection of SSWs. **SSWs** shall be inspected monthly, as a minimum, for compliance with 3.6.3, 3.6.4, 3.6.5, 3.6.7, 3.6.8, 3.6.9, 3.6.10, and 3.6.11.

4.4. Inspection of materials and equipment at SSWs. **SSWs** shall be inspected weekly for compliance with 3.6.6.

4.5. Inspection of operator disciplines. Inspection for compliance with 3.6.12 shall be conducted daily at 10%, minimum, of the **SSWs**, randomly selected.

4.6. Inspection of testing procedures. Inspection for compliance with 3.6.14 shall be conducted weekly.

4.7. Inspection of installation site practices. Inspection for compliance with 3.6.15 shall be conducted as often as practicable, preferably daily.

4.9. Inspection of packaging for delivery. Inspection for compliance with 3.8 shall be weekly.

4.10. Recertification of automated processes. Recertification of automated processes (3.9) shall be done monthly. (*Note:* Frequency of inspection depends on circumstances and is debatable. The above frequencies seem reasonable to us but are meant to be illustrative only.)

5. Notes

5.1. Definitions. Definitions of terms used herein are as follows:

Components: Electrical and electronic components.

Assemblies: Electrical and electronic subassemblies and assemblies up to but not including the unit level.

Units: Enclosed electrical and electronic higher assemblies.

Items: Components, assemblies, and units referred to collectively.

5.2. Precautions pertaining to ionized air blowers.

5.2.1 *Ozone generation.* All corona-discharge-type ionized air blowers generate ozone to some extent. The OSHA standard is 0.1 ppm for an 8-hour exposure. It is reported that 0.015 ppm ozone is "barely detectable by some sensitive people" while 1.0 ppm gives a "very noxious odor." Therefore, the rule of thumb is that if ozone can be smelled easily the concentration is too high for prolonged exposure.

5.2.2. *Shadowing effect.* Charged surfaces protected from the airstream by obstructions (the "shadowing effect") will not be neutralized effectively.

5.3. Precautions pertaining to antistatic polyethylene. This material is used for bags (typically 6-mil film), parts carriers, tote boxes, etc. It achieves its **antistatic** property by a weakly conductive "sweat layer" which is formed when an organic liquid antistat, added before extrusion, migrates or blooms to the surface and combines with water vapor from the ambient air. This invisible layer allows incipient **static charges** to bleed off before they can cause ESD damage. The antistat can cause contamination, and in the case of one product corrosion occurred. (*Note:* See Paper No. 9 in the Appendix.) The following considerations must be kept in mind when using **antistatic** polyethylene.

5.3.1. *Fading of pink color.* The identifying pink color will gradually fade to faint amber or off-white under fluorescent light or sunlight due to the ultraviolet portion of the spectrum. This fading does not affect the **antistatic** behavior; only the color changes.

5.3.2. *Cleaning.* The material should not be exposed to solvents. Water, alcohols, hydrocarbons, freons, or other solvents used in electronics manufacturing all remove the "sweat layer" and destroy the **antistatic** property, though new antistat will bloom to the surface and regenerate and sweat layer a few times until the internal reservoir of antistat is exhausted. The material must be cleaned only with an antistat solution, without drying; this procedure leaves an **antistatic** film on the surface. Other products such as premanently **antistatic** or **static-dissipative** film with a radiation-cured coating may also be solvent-sensitive and may lose their effectiveness after a single brief exposure.

(*Note:* Paragraphs 5.2 and 5.3 illustrate the kind of tutorial information that may be included in Section 5. Even though personnel have been taught these facts, caution notes are desirable to heighten the awareness of hazards that may damage hardware and, even more importantly, may endanger health or cause personal injury.)

Reference Document 1. General Design, Construction, and Maintenance Guidelines for ESD-Protected Areas

1. Introduction. This document gives general guidelines for **SSW**s in compliance with Model Specification 1. The term "ESD-protected area" means a location, e.g., a room or portion of a factory floor, containing a number of **SSW**s separated by aisles or other non-**SSW** zones.

2. SSW designation. The **SSW** shall be a clearly defined volume of space with caution signs, station identification numbering, and, when possible, borders marked with yellow tape. The "buffer zone" within the **SSW** but beyond the boundaries of the expected **SSZ** shall be as large as can be afforded in terms of space usage.

3. Housekeeping. The **SSW** shall be kept clean, orderly, and free of unnecessary **nonconductors** and unapproved materials and equipment.

4. Work surface. The **SSW** work surface shall be of an approved **static-dissipative** or **antistatic** material. **Grounding** of the work surface shall be certified annually in accordance with Reference Document 2, and the **SSW** shall have a current certification sticker.

5. Necessary nonconductors. Existing floors, utility rails, building structures, and standard seating (chairs and shop stools) may be tolerated within the volume in space of the **SSW** subject to the **CD Rule** in accordance with Model Specification 1. Plastic windows and fabric-covered walls are likely to violate the **CD Rule,** and even glass windows may be a problem. (*Note:* See Example 8 in Chapter 5.) If necessary, **conductive** chairs or chair covers may be procured from Reference Document 4.

6. Construction review. All plant layouts and construction design drawings shall be reviewed by Facilities and Industrial Engineering.

7. General guidelines. The following general guidelines shall be considered in planning and designing an ESD-protected area. These guidelines provide information beyond the scope of Model Specification 1, which is intended to give requirements with only a modicum of "how-to" detail. Thus, as new equipment becomes available, the present document will be updated to utilize it while Model Specification 1 will not need to be revised.

7.1. **Nonconductors.** Plain plastic or synthetic fabric (such as wall panels) shall be excluded if possible. Otherwise, necessary **nonconductors** shall be treated with **topical antistat** or kept at a safe distance from **ESDS items** by the **CD Rule.**

7.2. *Walls and other elevated surfaces.* Walls and utility rails constructed of gypsum board and painted with low-gloss latex are generally not an **apparent-charge** problem. Problems can come from plastic pipes, signs, windows, and special wall coverings such as high-gloss epoxy paint. The **CD Rule** shall be applied to these, and the opposite side of the wall shall be checked for power transformers, electric motors, or other machines which may produce **fields** that can penetrate the wall.

7.3. *Floors.* Tile or concrete floors must be treated with **static-limiting floor finish** in accordance with Model Specification 1. Floors with ordinary finish ("wax") or carpeting are an ESD hazard insofar as they fail to provide the desired safety net. Antistat-treated carpet, or carpet sold as **"antistatic,"** may be acceptable subject to the **CD Rule,** but ESD-protected areas should not be designed with carpet when concrete or tile would serve.

7.4. **Conductive** *floors.* These may be needed where the mobility of workers precludes the use of wrist straps, even with long cords and an overhead sliding trolley system. **Conductive** floors without **conductive** footwear or shoe straps are ineffective, and rubber-composition shoe soles may generate higher charges on the wearer when he walks on **conductive** floors than when he walks on **nonconductive** vinyl tile. Also, the resistance to ground may fall below the **ARTG** lower limit of 1 megohm (paragraph 3.6.5.3.2 of Model Specification 1) with wet leather shoes on a **conductive** floor, and we have recorded a case where the resistance was below 100,000 ohms so that a person touching a 110 V hot lead would have carried a possibly lethal current of more than 1.0 mA. This danger must be considered before buying and installing **conductive** floors and is a good reason, along with expense, to avoid them if possible. (*Note:* Our approach calls for ordinary tile with **static-limiting floor finish.**)

7.5. *Work surfaces.* Work surfaces must be an approved material from Reference Document 4. In areas where processes such as conformal coating, potting, or chemical processing are carried out, other work-surface materials such as stainless steel may be used *after* they have been added to Reference Document 4 with notes limiting their use to special situations. A **conductor** such as stainless steel is not a novel hazard at **SSWs**—hand tools, for instance, are often steel—but our philosophy is to minimize **conductive** surfaces and maximize **static-dissipative** and **antistatic** surfaces in order to best protect both personnel and **ESDS items.**

7.6. **Ioniziation. Ionization** can be used to reduce the ESD hazard in almost any area. However, the general use of **ionization** is not recommended because it will not allow any disciplines to be relaxed nor will it eliminate the requirement for any ESD-protective materials or permit the casual use of plain plastics. **Ionization** cannot limit the **triboelectric charging** process, as **humidification** can do, but does control the build-up of charges on **nonconductors**; thus the hazard of high **apparent charges** on some necessary items such as shields and laminar-flow booth side panels is reduced. In other words, standing charges are controlled by **ionization** of the room type, i.e., with overhead emitters. However, space charging could make room **ionization** part of the problem that it purports to solve, so this method should be considered only for special areas such as cleanrooms when the advantages are well-defined and the space-charging hazard is definitely controlled. Localized **ionization** is necessary in grit-blasting, spray-coating, and operation of some automated process machinery where **triboelectric charging** cannot be eliminated.

7.7. **Humidification.** Humidity control does limit the **triboelectric charging** process but does not eliminate any of the conventional safeguards; it is strictly a backup or "safety net" measure. Also, humidity control may give personnel a false sense of security and cause relaxation of **operator disciplines**, thus lowering ESD safety. Humidity control also is expensive and can cause corrosion or other adverse side effects. In conclusion, **humidification** is a backup that should be implemented only after careful consideration of benefits vs. cost and hazards.

7.8. *Lights and light fixtures.* Light fixtures with plastic covers or diffusers are potential hazards because of **fields** from the plastic as well as from the lighting element. Ultraviolet (UV) lights used in some inspection stations are safe when **ESDS items** are exposed for limited times as in normal inspection. (*Note:* See Example 27 in Chapter 5.)

7.9. *Maintenance.* ESD-protected areas shall be maintained under the close supervision of ESD-certified personnel. For example, only **static-limiting floor finish, antistatic** plastic trash-can liners, and approved cleaning agents must be used. No cleaning or janitorial work shall be done while **ESDS items** are out of their **Faraday cages** and not under the control of certified personnel.

(*Note:* This document is merely a model or sample and could be far more detailed in practice. Guidelines for special areas such as cleanrooms could run to many pages.)

Reference Document 2. Certification of Static-Safe Workstations (SSWs)

1. Introduction. This document gives procedures for certifying the **grounding** and the ground connections of certain critical, permanently fixed surfaces of **SSWs**.

2. Definition of SSW. Model Specification 1 defines an **SSW** as a volume in space which is designated and equipped to maintain an **SSZ,** which is a volume in space at every point of which the **Basic Rule** (paragraph 3.5.1 of Model Specification 1) is followed. The volume in space of the **SSW** includes assembly benches, ovens, desiccators, wave-solder machines, test equipment, and other equipment in which **ESDS items** are stored, processed, or handled.

3. Certification procedures.

3.1. *Certification decal.* Check that each **SSW** has a current certification decal to the requirements herein.

3.2. *Electrical resistance from work surface to ground.*

3.2.1. *Benches with* **static-dissipative** *or* **antistatic** *work surfaces with a buried* **conductive** *layer.* Measure the resistance from work surface to ground with a megohmmeter and an NFPA 99 electrode. Connect the negative ("low") lead of the megohmmeter to the electrode and the positive ("high") lead to the **grounding** lug. Place the electrode on the bench surface 18 plus or minus 2 inches from the lug. Measure the resistance from the electrode to the lug at 500 V. The resistance shall be between 1×10^5 and 1×10^{11} ohms. Alternatively, a Voyager RTG-210 Resistance to Ground meter or equivalent can be used. (When the resistance is under about 7 megohms using the palm of the hand as the electrode, a **Zapflash** will light. This is a useful quick check when the expected resistance to ground is toward the low end of the required range. If the **Zapflash** doesn't light, the resistance might still be acceptable, of course, by being in the middle or high end of the required range.)

3.2.2. *Benches with* **conductive** *work surfaces.* Using a multimeter at 0.5–50 V, probe between the work surface and the distribution wire (static ground). The resistance shall be 1 megohm, provided by a current-limiting resistor as done for **static-dissipative** or **antistatic** work surfaces. Laminar-flow booths, exhaust booths, and similar facilities where the **conductive** surface is electrically common with the structure shall have a resistance of 0–5 ohms by probing between the surface and commercial power ground. Note that **conductive** bench tops (work surfaces) are used in many clean-room and special-process applications where the nature of the task requires highly **conductive** materials such as stainless steel or aluminum.

3.2.3. *Processing, storage, and test equipment.* Using a multimeter at 0.5–50 V, probe between the commercial power ground connection on the chassis and bare metal on the front (or on the inside for ovens and desiccators) of the equipment. The resistance shall be 0–5 ohms. Verify the **grounding** of baffles in ovens and environmental chambers by probing in a similar way; again, the requirement is 0–5 ohms.

3.3. *Potential difference between static ground and commercial power ground.* Static ground shall be used for assembly benches (3.2.1 and 3.2.2). Other **SSW**s (3.2.3) may use commercial power ground. These two grounds shall not differ by more than 5 V dc or rms. **Continuous wrist-strap monitors** use both these grounds and will detect a gross failure in either.

3.4. *Wrist strap and ground wire connection.* Check that each **SSW** has wrist strap and ground wire connections in good condition. Alligator clips and twisted wire connections shall not be certified. All wrist straps must have **continuous wrist-strap monitors.**

3.5. *Resistance between* **grounding** *lug of the work surface and static ground distribution wire.* Check that this resistance is 1 megohm, provided by a series resistor in the lead wire which connects the **grounding** lug and the distribution wire. Make sure that the lead wire is a minimum of 18 AWG and the distribution wire is a minimum of 12 AWG. ("Minimum" refers to size; the higher the AWG, the smaller the conductor.)

3.6. *Resistance between static ground distribution wire and commercial power ground.* Check that this resistance is less than 25 ohms.

4. Identification of SSWs. Upon satisfactory completion of the above procedures, a certification decal shall be applied to the **SSW** to record compliance of **grounding** and ground connections with paragraph 3.6.5 of Model Specification 1.

Reference Document 3. Procurement Requirements for Continuous Wrist-Strap Monitor Unit

1. Scope. This document establishes requirements for a resistive (two-conductor) continuous wrist-strap monitor unit, including an alarm device and special wrist strap.

2. Applicable Documents. Model Specification 1, Handling and Assembly of Unpackaged **ESDS Items;** Reference Document 4, Approved ESD-Control Materials and Equipment.

3. Requirements.

3.1. *General requirements.*

3.1.1. *Overall design.* The unit shall consist of a monitor (alarm device) and a special wrist strap. The wrist strap shall consist of a two-snap wrist band and a two-conductor coil cord for connection to the monitor. The unit shall provide two points of skin contact and two ground connections. An alert shall be indicated in the event of an "open" (high resistance) or "short" (low resistance) in either circuit.

3.1.2. *Workmanship.* Workmanship shall equal or exceed industry standard, and fabrication shall be from good-quality components.

3.1.3. *Provision for calibration.* The monitor shall have provision for calibration and calibration seals.

3.1.4. *Documentation.* Service instructions, calibration procedures, and drawings, including a schematic of the circuitry, shall be provided by the supplier.

3.2. *Specific requirements.*

3.2.1. *Skin-to-ground resistance.* Total skin-to-ground resistance shall be continuously measured at 0.1–5.0 V and shall be greater than 1 ± 0.1 megohm and less than 10 ± 1.0 megohms with both circuits active.

3.2.2. *Wrist band.* The wrist band shall be (1) ¾-inch-wide Velcro with D-ring fastener or (2) adjustable, stretchable cloth. There shall be no **conductive** material in the band other than two metal skin contacts attached to the snaps. The contacts (plates or washers) shall be 0.1–0.6 inch apart. A wrist band shall be supplied as part of the standard unit. (Note: This is our preference based on tests for operator comfort. You may choose to allow stainless-steel expansion bands of the spidel type.)

3.2.3. *Coil cord.* The coil cord shall be resilient plastic, e.g., polyurethane or polypropylene, with flexible conductors and shall be a minimum of 6 feet in length when stretched to its practical limit. Banana plugs shall be provided for plugging into the monitor.

3.2.4. *Cord connection.* Cord connections shall be of the quick-disconnect spring type, which twists in any direction and allows release without damage to either the springs or banana plugs. This is a safety feature allowing operators to pull free easily.

3.2.5. *Power supply.* The monitor shall operate on standard 110 V, 60-Hz house power.

3.2.6. *Ground connections.* The system shall have two ground connections, one for each point of skin contact. One connection shall be to commercial power ground, e.g., via a three-pronged plug on the unit power supply, and the other connection shall be to a static-ground wire system in accordance with paragraph 3.6.5.2.3 of Model Specification 1.

3.2.7. *Alert signals.* Alert signals shall be (1) an amber light to indicate high resistance in either skin-to-ground circuit, (2) a green light to indicate that both circuits are in the specified resistance limits, (3) a red light to indicate low resistance (danger to operator), and (4) an audible alarm to accompany the amber or red light. (Note: You may choose to omit the low-resistance warning light.)

3.2.8. *Alert delay.* The monitor shall have an alert delay of 0.25–1.0 second to prevent annoying and unnecessary alert signals during a momentary discontinuity due to normal body movement.

3.2.9. *On-off switch.* The on-off switch, if any, shall be recessed or otherwise protected from being inadvertently turned on or off, e.g., by being struck by a knee.

3.2.10. *Provision for mounting.* The monitor shall have provisions for mounting either on top of or under the workbench surface.

3.2.11. *Safety.* The system shall be designed fail-safe for shock hazard.

4. Approved Source List. Units for purchase shall be selected from among those listed in Reference Document 4, paragraph 3.3.

Reference Document 4. Approved ESD-Control Materials and Equipment

(*Note:* This document lists materials and equipment selected according to the criteria of Reference Document 5 as well as other criteria such as cost. The list gives the name of the item, supplier's part number, supplier's location (city and state), and the basis for approval of the item. Approval may be by **standard test, special test,** or **analysis,** and the basis-for-approval column cites internal notebook page references, letters, and test reports, as well as suppliers' data. See Reference Document 5 for categories of materials and equipment listed.

This document is a "living list" which is constantly being added to and updated. Listing approved items directly in the specification has proved impractical in our experience because of the need for frequent revisions of the specification to keep the list current, each revision requiring a time-consuming coordination with the Engineering, Manufacturing, and Quality Assurance organizations.

Since ESD control is still a "young" field, novel products are always appearing. For example, a digital desoldering apparatus meeting all the requirements of DoD-STD-2000-1B was advertised in late 1988 as featuring "the only **static-dissipative** desoldering handpiece on the market." Since the need for treating a plastic handle with **topical antistat** would be avoided, this equipment is a good example of a candidate worth evaluating.

Of course, we can't list commercial products in this book, so now we'll proceed to Reference Document 5, which gives the criteria for choosing them.)

Reference Document 5. Criteria for Selection of Approved ESD-Control Materials and Equipment

Table of Contents

3.14. Metals
3.15. Test equipment
3.16. Miscellaneous

1. Introduction. The purpose of this document is to define criteria for selection of materials and equipment for handling unpackaged **ESDS items** in accordance with Model Specification 1. The materials and equipment which have been selected by these criteria are listed in Reference Document 4.

(*Note:* This document is merely illustrative. The criteria can and should be more extensive than those shown. Some will be **standard tests** cited by suppliers, and others will be **special tests** developed by the user.

Remember that many of the items listed are optional, as emphasized in some cases by asterisks and notes; this is not a "shopping list" of necessities but an all-inclusive compendium of items which might be used. In other words, the presence of an item on the list does not imply the desirability of the item for most applications. An example is room **ionization** systems, which we feel are applicable only to special areas such as cleanrooms *if* advantages such as an air-cleaning effect can be demonstrated and the space-charging hazard is controlled.)

2. Publications cited.

2.1. *Government documents.*

SPECIFICATIONS (Military):

MIL-B-81705	Barrier Materials, Flexible, Electrostatic-Free Heat Sealable

(Other government activity):

KCS-C-123	John F. Kennedy Space Center, NASA: Specification for Surface Cleanliness of Fluid Systems

STANDARDS

Federal

FED-STD-101	Preservation, Packaging, and Packing Materials: Test Procedures

Military

MIL-STD-1686A	Electrostatic Discharge Control Program for Protection of Electrical and Electronic Parts, Assemblies, and Equipment (Excluding Electrically Initiated Explosive Devices)

Department of Defense

DoD-HDBK-263	Electrostatic Discharge Control Handbook for Protection of Electrical and Electronic Parts, Assemblies, and Equipment (Excluding Electrically Initiated Explosive Devices)

2.2. *Non-government documents.*

SPECIFICATIONS

Model Specification 1	Handling and Assembly of Unpackaged **ESDS Items**

STANDARDS

American Society for Testing and Materials

Modified ASTM D257	Tests for DC Resistance or Conductance of Insulating Materials. (The modification is that a Voyager SRM-110 **Surface Resistivity** Meter or equivalent is used.)

National Fire Protection Association

NFPA 99	Standard for Health Care Facilities

OTHER PUBLICATIONS

Reference Document 3	Procurement Requirements for Continuous Wrist-Strap Monitor Unit

EOS/ESD Association and IIT Research Institute

EOS/ESD Symposia, 1979-current	Electrical Overstress/Electrostatic Discharge Symposium Proceedings (Annual)

Government-Industry Data Exchange Program (GIDEP)

GIDEP Alert MX-A-82-02 (March 21, 1983)	Manufacturing/Process Techniques, Wrist Strap, ESD Prevention
GIDEP Alert D5-A-84-01 (May 25, 1984)	Protective Clothing and Equipment, **Antistatic** Smock

(*Note:* Many other documents could be referenced. The ones given are merely illustrative.)

3. Materials/Equipment and Criteria. In the following subparagraphs, materials/equipment are listed in the left-hand column and criteria in the right-hand column.

3.1. *Bag materials and plastic films/sheets.*

a. Black, **conductive** carbon-loaded polyolefin	Volume resistivity greater than 1000 ohm-cm at low voltage, calculated from **surface resistivity** (ASTM D257) and thickness.
b. **Antistatic** polyethylene	Must be on QPL for MIL-B-81705, Type II. **Antistatic** agent must be a neutral, non-corrosive compound. Shelf life at R.T. for closed bags should exceed 8 years (EOS/ESD Symposium 1983; Paper No. 2 in the Appendix).

c. **Antistatic** clean nylon — Bleed-off less than 1 second for 5000V to 50V at 70°F and 50% R.H. Cleanliness level 50A by Kennedy Space Center KCS-C-123.

d. **Faraday-cage** foil laminate — All ESD-prevention requirements of MIL-B-81705, Type I, as well as an **antistatic** polyethylene liner with **surface resistivity** less than 10^{12} ohms/square. (MIL-B-81705 has an unsatisfactory liner.)

3.2. *Foam and bubble-wrap materials for cushioning.*

Antistatic foam and bubble-wrap — FED-STD-101, Method 4046: 2 seconds maximum charge decay time. **Surface resistivity** less than 10^{12} ohms/square.

3.3. *Apparel and operator* **grounding** *equipment.*

a. Smocks, **antistatic** — **Surface resistivity** less than 10^{12} ohms/square. (Successful smocks have been 6×10^6, 3×10^7, and 8×10^7 ohms/square.) Must not shed steel fibers (GIDEP Alert D5-A-84-01, 05-25-84).

b. Gloves and finger cots, **antistatic** — Less than 50 V **apparent charge** when stroked with polyester fabric. Fingertip wall thickness less than 0.001 inch. Resistant to inorganic acids and alkalis.

c. Wrist straps — Part of continuous monitor unit in accordance with Reference Document 3.

d. **Continuous wrist-strap monitor** — Gives alarm by light and sound when the **ARTG** exceeds 10 megohms, in accordance with Reference Document 3.

e. Heel strap, **conductive**	**Surface resistivity** less than 10^5 ohms/square.
f. Heel-grounder, with ankle strap	Resistance from heel-grounder to end of strap 1–2 megohms at 10 V.
g. Ground cord	Built-in 1-megohm resistor. Sturdy construction.

3.4. *Tote boxes, bins, trays, boxes, containers.*

a. **Static-dissipative** tote box	**Surface resistivity** at least 10^5 but less than 10^9 ohms/square
b. Portable bin system, **antistatic**	Less than 50 V **apparent charge** when rubbed with polyester fabric. Antistat must be internal (extruded into the plastic). **Surface resistivity** less than 10^{12} ohms/square.
c. Trays, **conductive** carbon-loaded polyolefin	**Surface resistivity** less than 10^5 ohms/square.
d. Box, transparent, **antistatic**	Less than 50 V **apparent charge** when rubbed with polyester fabric. Antistat must be internal (extruded into the plastic). **Surface resistivity** less than 10^{12} ohms/square.
e. Cardboard box, aluminum foil lined	When closed, must be electrically continuous over all internal surfaces.
f. Box, **conductive,** double-walled carbon-loaded polypropylene	**Surface resistivity** less than 10^5 ohms/square.
g. Cubical **antistatic** container	Less than 50 V **apparent charge** when rubbed with polyester fabric. Antistat must be internal (extruded into plastic). **Surface resistivity** less than 10^{12} ohms/square.

3.5. *Workbench tops and mats.*

a. Workbench tops (laminates)	Resistance less than 10^{11} ohms between **grounding** lug and a

	5-lb., 2.5-inch-diameter electrode (NFPA 99) on the surface. **Surface resistivity** less than 10^{12} ohms/square.
b. Table mat, soft, **static-dissipative**	Same as for workbench tops (laminates).

3.6. *Chairs, floor mats for chairs, seat covers.**

a. **Conductive** chairs	**Surface resistivity** of upholstery less than 10^5 ohms/square. When chair sits on **conductive** surface, less than 1 megohm resistance from upholstery to surface.
b. Hard **conductive** floor mats for **conductive** chairs	**Surface resistivity** less than 10^5 ohms/square.
c. **Conductive** seat covers	**Surface resistivity** less than 10^5 ohms/square.

3.7. *Floor materials (mats, tiles, carpets,* **static-limiting floor finish***).*

a. Floor mat, **conductive** or **static-dissipative***	**Surface resistivity** less than 10^9 ohms/square.
b. Floor tile, **conductive***	Electrical continuity between upper and lower surfaces.
c. Carpet, polyolefin, containing black, **conductive** fibers*	**Surface resistivity** less than 10^8 ohms/square (based on testing successful products). Electrical continuity between black fibers and **conductive** backing.
d. **Static-limiting floor finish**	Topaka slip 0.20 minimum, James slip 0.50 minimum, gloss 30 minimum, surface resistivity 10^{10} ohms/square maximum, when tested in accordance with the unabridged version of Paper No. 6 in the Appendix.

*Optional items.

3.8. *Static meters.*

Handheld **field meters**	Favorable review in the literature, e.g., EOS/ESD Symposium 1983 (Ref. 2-8). Easily calibrated by metrology.

3.9. *Air ionizing (***ionization***) equipment.*

a. Ionized air blowers	Neutralization rate for charged plastic: at 2 feet from the blower, time to reduce an **apparent charge** on polyethylene bubble-wrap from −3300 V to −2000 V shall not exceed 30 seconds. Minimum safe distance from corona-discharge points no more than 12 inches by **special test** with **MOSFETs** (see under "Air Ionizers" in Paper No. 3 in the Appendix).
b. Ionized air nozzles, electrical	Neutralization rate for charged surfaces: to be determined by experience. Minimum safe distance from corona-discharge points no more than 3 inches (Paper No. 3 in the Appendix).
c. Ionized air nozzle, nuclear	Neutralization rate for charged plastics: to be determined by experience. (There is little **field** from this equipment.)
d. Ionizer, pulsed, DC	Frequency must be at least 5 Hz (Example 21 in Chapter 5).
e. Grid, static-neutralizing	Minimum safe working distance (determined as for ionized air blowers) no more than 12 inches.
f. Room **ionization** systems*	**Apparent charge** on polyethyl-

*Optional item.

	ene film at bench level must decay from −5000 V to −500 V in less than 3 minutes; similarly, the **apparent charge** on polystyrene foam must decay from +5000 V to +500 V in less than 3 minutes. Isolated **conductors** at bench level must not be charged to more than 50 V. Balancing control must be automatic. Installation must be attractive in appearance.
3.10. *Desiccator cabinets.*	FR-4 circuit-board laminate must develop an **apparent charge** of no more than 300 V in normal operation (see Example 28 in Chapter 5). The cabinet must protect **ESDS items** from external **fields** as demonstrated by **analysis** or **special test.**
3.11. *Topical antistats.*	Chloride level less than 100 ppm for "chloride-free" type. Good wetting of polyethylene surface for chloride-containing type. Treated surface must be **antistatic.**
3.12. **Conductive** *tape.*	**Surface resistivity** less than 10^5 ohms/square.
3.13. *Shunt materials and lead-shorting devices*	
a. Foam, **antistatic** or **conductive,** shunt.	Must pass high-humidity test for corrosivity to device leads (see Example 14 in Chapter 5).
b. Connector shunt, **antistatic**	Made from material (Type II) on the QPL for MIL-B-81705. (*Note:* The user must satisfy himself that the feeble con-

	ductivity of an **antistatic** material has sufficient **shunting** effect; this is generally doubted.)
c. Connector shunt, **conductive,** carbon-loaded plastic	High-humidity exposure with gold-plated connector pins; see Example 14 in Chapter 5.
d. Lead-shorting devices	**Surface resistivity** less than 10^5 ohms/square. Volume resistivity less than 100 ohm-cm.
3.14. *Metals.*	Low electrical resistance.
3.15. *Test equipment.*	
Example: **Zapflash** multi-tester	Lights dimly at 7 megohms but not at 10 megohms or above.

(*Note:* This section includes **surface resistivity** meters, resistance-to-ground meters, megohmmeters, etc. See the check list of test equipment in Chapter 9.)

3.16. *Miscellaneous* (examples).

Voltage-suppressing screens for cathode ray tubes (CRTs) of computers or oscilloscopes	**Special test** for **apparent charge** using a **field meter.**
Scrub brushes	**Antistatic** or **conductive** handle. May have **nonconductive** plastic portion as allowed by **special test**; see Example 11 in Chapter 5.
Antistatic hand lotion	No lanolin or silicones. Acceptable **ARTG** when applied to dry skin.

Reference Document 6. Certification of Personnel for ESD Control

In accordance with paragraph 3.6.13 of Model Specification 1, personnel shall receive a minimum of 8 hours of training and pass a test for certification, and they shall be recertified annually by taking a minimum of 2 hours of brush-up training and passing a test.

The training course shall emphasize (1) compliance with Model Specification 1, (2) "how-to" procedures, and (3) the "why" of procedures and rules being followed.

(1) Experience has shown that training course material and the specification being used may be poorly related and even contradictory! The objective should be to teach people to comply with and implement the specification, not to question it, ignore it, or violate it because training gave them other ideas.

(2) Among "how-to" procedures, **operator disciplines** should be heavily stressed. Remember that an ESD "guru" could do assembly work safely at a wooden table with no wrist strap, because knowledge and techniques are more important than the best ESD-control materials and the most expensive, sophisticated equipment.

(3) The logical derivation of all ESD-control procedures from the **Basic Rule** is the essence of our "A-to-Z" method, and operators and other non-engineers must be given a simple conceptual picture of ESD damage models that makes every precaution understandable. **Fields** may be pictured as a "toxic cloud" whose diminishing density is judged by **apparent charge** and distance, so that the principle of the **CD Rule** is made obvious. **Discharges** are easily understood as sparks or else subtle, invisible "sparks" that can "zap" a device; thus an operator will see the imperativeness of **grounding conductors** to maintain the same potential and avoid the flow of electrons like a liquid from a higher level to a lower. Electrons can be pictured as "static dust" that rubs off of one surface and contaminates another during **triboelectric charging.** Concepts like the **ARTG** are a little more difficult to explain; in this case, a momentary surge of voltage must be envisioned, requiring limited resistance (enough conductivity) to drain off the electrons so quickly that they can't accumulate to the danger level.

Visual aids such as video tapes are valuable *if* they are tailor-made to fit the specification. A demonstration with a curve-tracer, showing a **MOSFET** being damaged by an ungrounded person's touch, is a vivid demonstration of proven value in convincing the audience that ESD damage really happens. We remember one video tape which showed an airplane crashing in a ball of flame because of **latent failure.** This may be a bit exaggerated, but horror stories of ESD destruction of expensive assemblies (see Example 25 in Chapter 5) are documented and cogent. No employee would want to be a guilty party in such an infraction, and no one needs to be if he follows the rule; "If there's doubt about ESD protection in this process, ask Engineering for advice." As in any field, the operator or engineer on the shop floor must know the limits of his knowledge so that he won't overstep them and risk either overt damage or the undetected "wounding" of devices that can cause those hard-to-prove but always possible **latent failures.**

A suitable test is 50 or 100 questions with multiple-choice answers and a passing grade of, say, 80%. The test should reveal whether the student understands how precautions and rules flow from and implement the **Basic Rule.** People tend to follow instructions when good reasons are given but to ignore "senseless" orders.

We have found that decals applied to identification badges are useful in screening unqualified people from ESD-protected areas. Also, the decals raise the ESD-awareness of their wearers and, like miniature diplomas, are a source of pride.

Thorough records of certification and recertification are required by MIL-STD-1686A and should be kept in any event. If hardware should be mishandled, the testimony of certified personnel will be more credible when their training is documented. For example, an operator might say, "When my wrist strap pulled loose, I was so absorbed in my work, which involved touching **ESDS** leads sensitive to 300 V, that I ignored the alarm of the continuous monitor for a minute. But as soon as I realized what had happened I measured my voltage, by pointing a **field meter** at the grounded work surface, and found it to be less than 100 V even when I shuffled my feet (on **static-limiting floor finish**), which I hadn't been doing during the infraction." Based on this report by an astute and certified person, if the hardware tested O.K. it could be assumed to be safe from **latent failure.**

We emphasize once again that an operator is analogous to an airplane pilot who can either "fly" or "crash" the most costly, well-equipped **SSW.** Since the operator's skills are vital, he or she should have considerable autonomy and, as Dan Anderson says, should be "vice-president" of the **SSW.** An authoritarian management system which withholds information because "knowledge is power" will not only keep operators ignorant but make them indifferent and even hostile. In ESD control, personnel must be trusted as well as trained.

Model Specification 2. Packaging of **ESDS Items** for Shipment, Storage, and In-Plant Transfer

1. Scope

This specification establishes requirements for packaging constructions which meet the definition (3.2, below) for a static-safe package **(SSP).** An **SSP** protects the packaged **ESDS item** from both **fields** and **discharges.**

2. Applicable Documents

The following documents, of the latest issue in effect except as otherwise indicated, form a part of this specification to the extent specified herein. In the event of conflict between documents referenced herein and the contents of this specification, the contents of this specification shall govern.

2.1. Government Documents.
SPECIFICATIONS

Federal

RR-W-365	Wire Fabric (Insect Screening)
PPP-C-795	Cushioning Material, Packaging (Flexible Cellular, Plastic Film) for Packaging Applications
PPP-C-1752	Cushioning Material, Unicellular Polyethylene Foam, Flexible (for Packaging Purposes)
PPP-C-1797	Cushioning Material, Resilient, Low Density, Unicellular, Polypropylene Foam
PPP-C-1842	Cushioning Material, Plastic; Open Cell for Packaging Applications

Military

MIL-B-131	Barrier Materials, Waterproof, Greaseproof, Flexible, Heat-Sealable
MIL-S-19491	Semiconductor Devices, Packaging of
MIL-M-38510	Microcircuits, General Specification for
MIL-R-39032	Resistors, Packaging of
MIL-B-81705	Barrier Materials, Flexible, Electrostatic-Free, Heat-Sealable
MIL-P-81997	Pouches, Cushioned, Flexible, Electrostatic-Free, Reclose-able, Transparent

STANDARDS
Military

MIL-STD-129	Marking for Shipment and Storage
MIL-STD-280	Definitions of Item Levels, Item Exchangeability, Models, and Related Terms

MIL-STD-794	Parts and Equipment, Procedures for Packaging of
MIL-STD-1285	Marking of Electrical and Electronic Parts
MIL-STD-1686A	Electrostatic Discharge Control Program for Protection of Electrical and Electronic Parts, Assemblies, and Equipment (Excluding Electrically Initiated Explosive Devices)

2.2. Non-government documents.
SPECIFICATIONS

Model Specification 1	Handling and Assembly of Unpackaged **ESDS Items**

STANDARDS
American Society for Testing and Materials

ASTM D257	Tests for DC Resistance or Conductance of Insulating Materials
ASTM D999	Vibration Testing of Shipping Containers

3. Requirements

3.1. Order of precedence. In the event of conflict between the requirements of this specification, a program contract, and drawings calling out this specification, the requirements of the contract and the drawings shall take precedence in that order.

3.2. Static-Safe Package (SSP). Packaging shall achieve an **SSP,** which is defined as an **SSZ** (3.2.1) within a **Faraday cage** (3.2.2) or within material providing sufficient **shielding** to pass a **discharge test** (4.3.3) on the particular configuration of shipped/stored **ESDS item** and packaging. The **Basic Rule** (3.2.3) is assumed to be followed in regard to **triboelectric charging** when **antistatic** materials known to limit charging by the "lubricity effect" are used as the liner. (*Note:* This exemplifies the kind of assumptions necessarily made in any specification for ESD control. **Special tests** for every combination of **ESDS item** and package would give high confidence in protection but are impractical.)

3.2.1. *Static-Safe Zone (***SSZ***).* An **SSZ** shall always be maintained around an **ESDS item,** with an **SSZ** being defined as a volume in space at every point of which the **Basic Rule** (3.2.3) is followed.

3.2.2. **Faraday cage.** The **Faraday cage** which is part of an **SSP** (3.2) is defined as a container whose wall is electrically continuous and has at least the electrical conductivity of 0.00025-inch (0.25-mil) aluminum foil. Small holes, such as made by staples, are allowed.

3.2.3. **Basic Rule. ESDS items** shall always be protected by the **Basic Rule,** which is: An **ESDS item** shall never, even for a nanosecond, be exposed to an **E field** in violation of the **CD Rule** (paragraph 3.6.11 of Model Specification 1) or have its sensitive leads or terminals touched to, or receive a **discharge** from, any surface at more than 50 V.

3.3. Materials. The following materials shall be used subject to the requirements of this specification.

3.3.1. *Specified materials.*

a. Bag, foil laminate, **antistatic** version of MIL-B-131, Class 1
b. Bag, MIL-B-81705, Type I
c. Bag or film wrap, MIL-B-81705, Type II
d. Pouch, MIL-P-81997, Type II
e. Cushioning, PPP-C-1752, Type VII, Class 4
f. Cushioning, PPP-C-1842, Type VIII, Style A or B
g. Cushioning, PPP-C-1797, Type II
h. Cushioning, PPP-C-795, Class 2
i. Wire fabric, RR-W-365, Type VII, 18 × 16 regular
j. Bag, laminate with see-through metallization covered by an **antistatic** film (exposed metallization, or metallization protected only by a coating instead of a film, is undesirable)
k. Bag, cellulosic, transparent, permanently antistatic, noncorrosive, for intimate wrap

3.3.2. *Other materials.* The materials shall conform to the following requirements.

3.3.2.1. **Conductive** *material.* Noncorrosive **conductive** ESD-protective material shall have a **surface resistivity** of less than 10^5 ohms/square.

3.3.2.2. **Static-dissipative** *material.* **Static-dissipative** material shall have a **surface resistivity** of at least 10^5 ohms/square but less than 10^9 ohms/square when measured at 70°F and 50% relative humidity.

3.3.2.3. **Antistatic** *material.* **Antistatic** material shall have a **surface resistivity** of at least 10^9 ohms/square but less than 10^{12} ohms/square when measured at 70°F and 50% relative humidity.

3.3.2.4. **Nonconductive** *material.* **Nonconductive** material is defined as having a **surface resistivity** of at least 10^{12} ohms/square when measured at 70°F and 50% relative humidity, minimum.

3.4. Conformance of package to an SSP (3.2). The package shall provide **Faraday-cage** or equivalent protection(3.4.1) and shall limit **triboelectric charging** (3.4.2) within its interior.

3.4.1. **Faraday-cage** *or equivalent protection.* Protection from **fields** and **discharges** shall be provided by materials conforming to the definition of a **Faraday cage** (3.2.2), e.g., materials 3.3.1.a, b, or i, or else the **discharge test** of 4.3.3 shall be passed. (*Note:* In practice it may be necessary to allow the use of 3.3.1.j bags to accommodate suppliers, and these bags may pass option 1 of paragraph 4.3.3 but not option 2. However, we discourgage the use of option 1 because 100 pF is low for people and 1500 ohms is high; 185 pF and 150 ohms as in option 2 provide a realistic worst case. See "**HBM**" in Chapter 2.)

3.4.2. *Limiting of* **triboelectric charging.** Because of lubricity (3.2), the materials in 3.3.1, except b and i, are satisfactory. Other materials (3.3.2) shall pass the **triboelectric charging** test of 4.3.4, or else the **ESDS item**(s) in the package shall be mounted securely so that sliding cannot occur. Even with approved liner materials, sliding shall be minimized, e.g., by the pressure of cushioning material. (*Note:* Material 3.3.1.b has a **nonconductive** liner which caused charging that damaged **MOSFETs** in a published test (Ref. 2–7). However, this bag may be lined with a bag of 3.3.1.c or 3.3.1.k in the technique of "double-bagging" (Ref. 6–1 and "Conclusions on Bag Materials" in Paper No. 1 in the Appendix). Alternatively, the packaged **ESDS item** could be prevented from sliding inside the 3.3.1.b bag.)

3.4.3. **Surface resistivity.** Both interior and exterior surfaces shall be **conductive, static-dissipative,** or **antistatic,** or they may be **nonconductive** if the **apparent charge** never exceeds 50 V because of **voltage suppression.** Examples of the latter are carrying cases made of anodized aluminum or painted steel. Other examples are foil bags: 3.3.1.b or MIL-B-131, Class 1, which is **nonconductive** both inside and out. However, the safest policy is to double-bag with 3.3.1.b (see Note in 3.4.2) or to use the **antistatic** version of MIL-B-131, Class 1, which is 3.3.1.a.

3.5. Corrosivity. The interior of the package shall be noncorrosive when tested in accordance with 4.3.6.

3.6. Contamination. The interior of the package shall be noncontaminating when tested in accordance with 4.3.7.

3.7. Rails and carriers. Rails shall be **conductive, static-dissipative,** or **antistatic** and shall meet the requirements of 3.4.2, 3.5, and 3.6. For supplier packaging, carriers may be **nonconductive.** (*Note:* **Nonconductive** carriers are allowed for two reasons. First, their small suface area limits their **apparent charge** and their **field.** Second, their acceptance is a matter of practicality because so many devices are provided in them.)

3.8. Shunting. **ESDS** leads, terminals, or contacts shall be protected by **shunting** whenever possible.

3.9. Cardboard. Glazed cardboard shall not be used unless the glaze is a special **antistatic** composition.

3.10. Physical protection. **ESDS items** shall be packaged in a manner that will prevent damage and contamination (3.6) during handling and transportation.

3.11. Cleanliness. Prior to packaging, **ESDS items** shall be clean and free from foreign material in accordance with the applicable component or assembly specification.

3.12. Uniformity. Containers shall be uniform in size and shape for shipment but not necessarily for storage or in-plant transfer.

3.13. Closure. Closure shall be secure, as needed, and according to instructions.

3.14. Marking. Marking shall be in accordance with MIL-STD-129. Also, MIL-STD-1285 will be applicable as tailored for individual programs.

3.14.1. *Description of contents.* Each package shall be marked with the following information as a minimum:

a. Part No. or other identifying number of **ESDS items** inside package
b. Supplier's trademark or name
c. Quantity in package
d. Lot or date code (when applicable)

3.14.2. *Transparency.* Transparent containers need not be marked when markings inside can be read easily. (*Note:* For transparency as well as superlative **shielding,** use 3.3.1.i sandwiched between layers of 3.3.1.c.)

3.14.3. *Marking for ESD-sensitivity*

3.14.3.1. *Exterior packs.* Exterior packs shall be marked with a yellow caution label having black lettering. The MIL-STD-1285 symbol (beneath this paragraph) shall be used, and the lettering shall say: "CAUTION NOTE: This equipment contains parts and assemblies sensitive to damage by electrostatic discharge (ESD). Use ESD precautionary procedures when touching, removing, or inserting." A 2 × 2 inch label shall be placed on one side of each intermediate container. Two 4 × 4 inch labels shall be placed on each exterior container—one on the identification marking side (or surface) and one on the opposite side of each shipping container exceeding ½ cubic foot. Smaller shipping containers shall be marked in the same manner except that the 2 × 2 inch label shall be used as an alternate to the larger one.

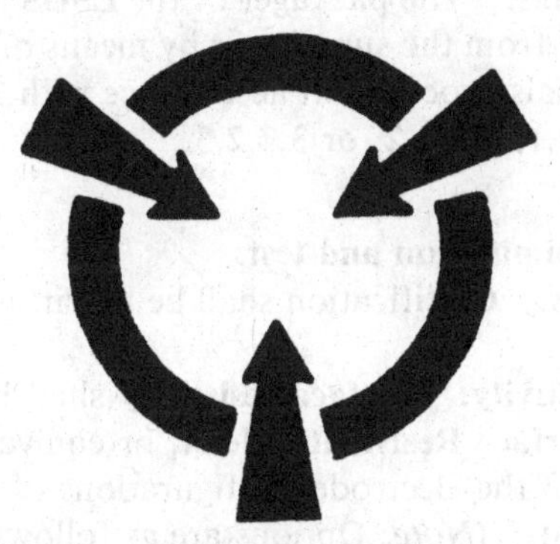

3.14.3.2 *Interior (unit) packs and containers for storage or in-plant transfer.* Unit packs or containers for storage or in-plant transfer shall be marked with the MIL-STD-1285 symbol (above), or the EIA symbol (paragraph 3.6.16.2 of Model Specification 1), and the statement: "Do Not Open Except at a Static-Safe Workstation." Minimum size of the symbol shall be one-third of an inch measured vertically. The symbol shall be printed in black or in the same color as the identification marking if this is other than black. When available marking space permits, the caution note of 3.14.3.1 may be used.

3.15. Other packaging requirements. (*Note:* Many special requirements are possible, especially for physical protection, but the purpose of this model specification is to emphasize ESD-protection requirements. On military contracts, documents such as MIL-STD-794, MIL-S-19491, MIL-M-38510, and MIL-R-39032 may be cited.)

3.16. Containers for storage or in-plant transfer. It is emphasized that containers, such as tote boxes or bins, for storage or in-plant transfer of **ESDS items** shall meet all the requirements of this specification.

4. Quality Assurance Provisions

4.1. Responsibility for inspection. Unless otherwise specified in the contract or order, the packager of the **ESDS items** shall be responsible for the performance of all testing required herein except for first-article inspections which are not quality-conformance inspections; the buyer shall be responsible for the latter.

4.2. Materials of construction of package.

4.2.1. *Specified materials.* The packager of the **ESDS items** shall verify by means of certification from the material supplier that materials procured in accordance with 3.3.1 are in compliance with the specifications.

4.2.2. *Other materials.* The packager of the **ESDS items** shall verify by means of certification from the supplier or by means of tests in accordance with 4.3.2 that materials procured in accordance with 3.3.2 are in compliance with either 3.3.2.1, 3.3.2.2, or 3.3.2.3.

4.3. Methods of examination and test.

4.3.1. *Certifications.* Certification shall be by virtue of acceptance and shipment of the order.

4.3.2. **Surface resistivity. Surface resistivity** shall be measured with a Voyager SRM-110 **Surface Resistivity** Meter, or equivalent, in accordance with the intent but not the electrode configurations of ASTM D257.

4.3.3. **Discharge test.** (*Note:* Options are as follows:

(1) *High-resistance* **discharge test.** The test apparatus is composed of (1) a steel probe with resistor and capacitor and (2) a pair of disks to be placed inside the package. The package shall be arranged so that both disks are pressed tightly against its inside surface. The probe is $3/8 \pm 1/16$ inch in diameter and 3 ± 0.25 inches long, and it has a hemispherical tip. It is connected by a 45-inch-long, 16-AWG insulated lead to a 1500-ohm resistor, which is connected by a 4-inch-long, 16-AWG lead to a 100-pF capacitor. The capacitor is charged at 10,000 V (positive, but either polarity might be used). Then, starting at 12 ± 2 inches from the package, the probe is moved toward the package at a speed of 5 ± 2 feet/second until contact is made at the place where one of the disks (metallic, 0.75 inch diameter, 0.050–0.10 inch thick) touches the inside of the package. This upper disk is connected to the gate lead of a 2N4351 or equivalent **MOSFET** with an **HBM** rating of about 100 V, and the lower disk is connected to the substrate-case lead. The package rests on a grounded metal plate so that the lower disk is separated from the plate by the wall of the package. The disks are joined in a "capacitive sensor" which separates them by 0.5 inch of **nonconductive** plastic (see Fig. 2 of Paper No. 3 in the Appendix). Alternatively, the disks

may be separated from each other by a minimum of ½ inch of air inside the package. Five consecutive **MOSFETs** are tested, and protection is judged inadequate if one or more of the devices shows a shift of more than 0.04 V in the gate-source threshold voltage ($V_{GS(TH)}$), if the device is shorted, or if the current-voltage curve has changed in shape. Data are obtained with a curve-tracer, Tektronix Model 576 or equivalent.

(2) *Low-resistance* **discharge test.** The test conditions and criteria for protection are exactly the same as for the above high-resistance test except that the capacitance is 185 pF and the resistor is 150 ohms. This is Test 5 in Table 2 of Paper No. 4 in the Appendix. We advise using this option unless expedience necessitates option 1. A variation on this test using square electrodes attached to the inside of the walls is suitable for tote boxes; see Paper No. 7 in the Appendix.)

4.3.4. **Triboelectric charging** *test.* (*Note:* A **standard test** using quartz and Teflon (Appendix G of EIA 541) is available, but the reader is advised to develop his own test with the most common surfaces he plans to package. For example, FR-4 circuit-board material was used in Papers No. 2 and 7 in the Appendix, and aluminum and an acrylic coating were also tested in Paper No. 7. Use a **Faraday cup,** and either shake coupons in a bag (as in the EIA 541 test), shake a box lined with the material in question so that coupons slide back and forth over it (Paper No. 7 in the Appendix), or use the roller test developed by 3M and applied to floor finishes in Paper No. 6 in the Appendix. The allowable limit of charging expressed in nC (nanocoulombs) must be set based on experience.)

4.3.5. **Shunting.** The item shall be visually inspected, with magnification as required, for proper **shunting.**

4.3.6. *Corrosivity.* (*Note:* The user must select or develop appropriate tests for his packaged items. A possibility is TAPPI T-406 (not referenced in this specification) for tarnishing of silver, solder, and copper by reducible sulfur. Another is a high-humidity, elevated-temperature test as used to indicate galvanic corrosion in Example 14 in Chapter 5 or to detect corrosion of solder by transferred antistat in Paper No. 9 in the Appendix. The tests must be realistic so that corrosion will be avoided but cost-effective packaging materials won't be ruled out; in contrast, unrealistic laboratory tests might either pass an inadequate material or require such unnecessarily high performance that a cost-effective candidate could fail.)

4.3.7. *Contamination.* (*Note:* Again, the user must select or develop pertinent, realistic tests. See Paper No. 9 in the Appendix for potential contamination problems from transferred antistats. Other materials may also cause trouble, e.g., carbon-loaded polyolefin which sloughs **conductive** particles or a fabric or other composition containing stainless-steel fibers which might shed a fiber to short out two circuit lines of a printed wiring board not yet conformally coated.)

4.3.8. *Physical protection.* The packaged item shall be visually examined with the unaided eye for damage to the item as well as damage to the package. When required, the completed package shall be subjected to vibration testing in accordance with ASTM D999, after which the **ESDS item** shall be functionally tested in accordance with the applicable specification for item performance.

4.3.9. *Cleanliness.* The item shall be visually examined to determine that required cleanliness has been accomplished.

4.3.10. *Uniformity.* Packages shall be visually examined for adequate uniformity.

4.3.11. *Closure.* The closure shall be visually examined.

4.3.12. *Prescribed packaging.* The package shall be examined and tested to determine that all elements of the prescribed construction have been used and that cardboard and paper are unglazed or have a special **antistatic** glaze. Foil thickness shall be verified with a micrometer.

4.3.13. *Marking.* Visual examination of the completed package shall verify the proper application of the specified marking.

5. Notes

5.1. EMI/RFI shielding. EMI/RFI **shielding** is outside the scope of this model specification. However, protection is afforded by foil or screen, e.g., 3.3.1.a or 3.3.1.i.

5.2. Magnetic shielding. **Shielding** from *H* fields is outside the scope of this specification. Iron foil may be used.

5.3. Definitions. MIL-STD-280 defines such terms as "item levels" and "models."

Chapter 7

Program Organization and Implementation

This chapter includes several topics: reasons for an ESD-control program; history of ESD control and purpose of this chapter; objectives and overview of program; what, when, and how; cost-effectiveness; plan of action; the enforcement problem; advice to small companies; the role of the program coordinator; and safety.

Reasons for an ESD-Control Program

One reason for an ESD-control program is that it is required by a Government contract. In this case, compliance with MIL-STD-1686A (formerly DoD-STD-1686) will be necessary, and three Data Item Descriptions (DIDs) are involved: DI-RELI-80669 for an ESD-control program plan, DI-RELI-80670 for reporting results of ESD-sensitivity tests of electrical and electronic parts, and DI-RELI-80671 for handling procedures for **ESDS items.** These will be Contract Data Requirements List (CDRL) items to be submitted for approval by the procuring agency.

A second reason is that an effective ESD-control program saves money by lowering rejection rates of products. Many articles and papers have attested to large savings. For example, one company spent $200,000 on ESD control beginning in 1980. Before that, ESD damage was estimated at $800,000 to $1,200,000 per year; afterwards, it fell to under $100,000 per year. Also, warranty repairs fell from $270,000 per year to $180,000 per year (Ref. 7-1). Another large corporation achieved savings in remanufacturing and repair that were anticipated to reach $21,000,000 in 5 years (Ref. 7-2). Incidentally, selection of the right materials and equipment can save much money within the ESD-control program itself. Ref. 7-2 cites the purchase of wrist straps and charge-draining materials costing about $6000 instead of $50,000 worth of air-ionizing blowers, which would have been inadequate by themselves anyway. We'll say more about prudent expenditures later in this chapter.

A third reason for an ESD-control program is to prevent **latent failures,** although instances of these are difficult to prove (see **"Latent Failure"** in Chapter 2). The possibility of **latent failure** could be enough to cause mishandled hardware to be declared suspect or even quite unacceptable by a Government customer after an unfavorable audit of manufacturing practices. MIL-STD-1686A is sketchy enough for people to rationalize away the requirements in a token effort, but the Department of Defense is beginning to realize that **latent failures** are possible in mission-critical equipment. Therefore, the war against ESD damage must be fought even though the enemy is invisible—a process that might be likened to exorcism, but with ghosts which could be real. Don't let disciplines become lax with the excuse that yields are high. Besides yields, the reliability of hardware is at stake, and deficiencies can become evident only over the long term.

Remember that every ESD event does not cause measurable damage to hardware. In fact, many ESD experts have concluded that 90% of ESD damage is less than catastrophic and will not be detected until the degradation develops into a full functional failure, or intermittent malfunction, after the product is in use. These devices are walking wounded or candidates for **latent failure,** and they may cause unnecessary rework/retest hours, field returns, and a decrease in MTBF and MTBR. Minimizing the cost of dealing with the walking wounded will put an effective ESD-control program on a paying basis, though savings aren't obvious as they are for products killed by ESD during manufacturing.

History of ESD Control and Purpose of This Chapter

During the past several years, ESD-control measures have been implemented in most major electronics manufacturing firms. At first a few simple controls were imposed on assembly operators. Then the addition of more complex procedures caused turmoil while the organizations adjusted. As time passed, uncongenial additions were eliminated or ignored until they disappeared. With the advent of DoD-STD-1686, the industry was challenged with converting existing control measures into formal programs. The commercial sector, which would ordinarily not be concerned with Government documents, is finding this Standard (now MIL-STD-1686A) useful as a guide for establishing a technically sound, cost-effective ESD-control program. Unfortunately, the measures used by many organizations have grown in an unplanned way. The purpose of this chapter is to suggest a systematic approach to formalizing a program with flexibility to grow with ever-changing ESD-control technology and products. Readers who already have formalized programs will find suggestions for reevaluating and upgrading program elements.

Objectives and Overview of Program

Simply stated, the objective of an ESD-control program is to protect **ESDS** hardware from damage by ESD. This requires a coordinated effort by every function from the Chief Executive to the new hire in the floor-maintenance crew. All too often the ESD-control program is viewed as a set of rules inflicted on assemblers and technicians who process hardware. This distorted viewpoint destines a program to fail, in the sense that the program costs more than it saves. Remember that operators stand in the front line of the ESD battle, but without the logistics of the rear echelon providing materials, equipment, and direction, the battle will be lost. Therefore, a total management system must be developed to support the front lines.

A realistic and systematic approach is needed. This requires information on the flow of product through the plant, from receipt and inspection of raw materials to quality control and shipment of finished hardware. The factory is a complex, delicately balanced structure whose operation is almost impossible to diagram because many support functions are not clearly defined. However, overlooking one seemingly obscure function can jeopardize an ESD-control program. Engineering organizations may be equally bewildering, but none of their elements can be slighted. Only a total, integrated effort can succeed.

The beginning of the development of an ESD-control program is an information-gathering process in which the following questions must be answered:

1. What is the lowest *V*-zap level (see "**HBM**" in Chapter 2) in each process?
2. What control methods are best suited to each step in the process? For example, are wrist straps preferable to **conductive** floors combined with **conductive** footwear? (We recommend wrist straps, with long cords and overhead trolleys if necessary, whenever possible. **Conductive** floors can be hazardous as pointed out in Reference Document 1 in Chapter 6.)
3. What level of personnel training will be required?
4. Are existing policies supportive of an ESD-control program?
5. What new company policies will be required to enforce ESD-control disciplines?
6. What special skills will be required in each area, e.g., assembly versus testing?

More questions will become apparent as the information-gathering process proceeds.

Now comes the real challenge: arranging the information in a meaningful format. One practical method is to construct a functional profile delineating the interaction of each function with other activities. This profile should identify key individuals in the management structure. The profile should include a matrix tabulating elements such as workstation types, worker skills, special process requirements, environmental conditions and requirements, and other conditions impacting ESD-control disciplines.

What, When, and How

What, when, and how are the questions for program implementation. What areas and processes require ESD control will be easily identified from the functional profile. "How" is the tough question, so let's look at "when" first. It is impractical to implement total ESD control all at once. The first control measures should be introduced concurrently with personnel training in the areas of most vulnerability to ESD. These initial measures are easiest to sell because their cost-effectiveness is readily demonstrated. After the first disciplines are in place, refinements and expansion of ESD controls to the balance of the factory will commence. With refinements, diligent attention to detail will be needed to prevent reverting back to the old way. Workers will have to break old habits and establish new ones, and management's failure to recognize and support the habit-changing aspects of the program will cause it to be ineffective and fail to produce the desired cost savings.

"How" involves the most controversy. Each salesperson selling ESD-protective materials and equipment claims to have the best "how" and will quickly tell you why competitors' "hows" will not work. The only practical solution to this dilemma is to buy only products that clearly implement the **Basic Rule.** Remember that three main damage mechanisms are at work. First is **DI,** of which the **HBM** is the imperfect simulation (too high resistance, for one thing) of the most common source of current flow: touching by people. Second is the **CDM,** in which the device is charged and subsequently is damaged by **DI** when an **ESDS** lead touches a surface with some conductivity, especially a truly **conductive** one. Third is the **FIM,** in which a **field** causes a difference in potential to be induced in the device so that the device "zaps" itself from within. A special case of this is the **FFB,** in which a **field** develops on the board holding the devices; when the damage occurs during subsequent **discharge,** the **FFB** resembles the **CDM.** The **FIM** is the most insidious of the damage models because of the complex nature of internal circuitry and the unknown locus of electrical energy imparted by a **field.** Repeatability in **FIM** testing is difficult, and data can't predict the outcome of a subsequent test where one of the parameters such as **field strength** is changed. There is no **standard test** established by Government or industry for establishing reliable tolerance thresholds. However, a conservative safe-distance range has been empirically developed and over the years has proven safe and reliable; this is our own **CD Rule.** With the above three damage models in mind, materials and equipment should be selected to combat specific threats. For example, **antistatic** materials, as opposed to **conductive,** will be chosen to discourage the **CDM** by lessening the rapidity of **discharges.**

Cost-Effectiveness

Special ESD-control program design objectives should be developed around facilities and equipment as configured in the shops and laboratories engaged in producing **ESDS** hardware. Therefore, no one program will fit all occasions. Cost-effectiveness, which is the driving force with top management, is attained by accurately defining control requirements and fitting them to existing conditions. Factors in selecting these requirements are: (1) lowest level of ESD-sensitivity, (2) mission criticality of the product, (3) cost of **ESDS** components, (4) cost and feasibility of rework, (5) schedule impact of ESD failures, and (6) willingness and ability of workers to accept and implement more or less complex disciplines of ESD control.

In this book we have made the following assumptions about the program illustrated by our model specifications in Chapter 6:

1. Lowest level of sensitivity: 100 V as determined by the **HBM,** corresponding to a maximum safe **apparent charge** of 300 V on **nonconductors.**
2. Mission criticality: high , e.g., a life-support system.
3. Cost of sensitive components **(ESDS items)**: moderate to high.
4. Cost and feasibility of rework: cost moderate to high, feasibility varied.
5. Schedule impact of failures requiring rework: manageable.
6. Worker willingness and ability: good (but may require union cooperation).

Operator disciplines are the cornerstone of a cost-effective program emphasizing skillful handling rather than an attempt to "buy one's way out of the problem" by "throwing money at it." Indeed, such an attempt would fail, because the most expensive materials and equipment are useless if not used intelligently. Some key **operator disciplines** (see paragraph 3.6.12 of Model Specification 1 in Chapter 6) are:

1. Never touch an **ESDS** device unless absolutely necessary.
2. When a device must be touched, avoid the leads and handle it by the case.
3. Touch devices only to approved materials which are grounded. **Antistatic** materials are safer to touch than are **conductive** materials (Papers No. 3 and 7 in the Appendix).
4. Never subject a device to a **field** of an unknown safe level. The safe level (safe distance for a given **apparent charge**) is determined by the **CD Rule.**
5. Before a device leaves the **SSW,** place the device in an **SSP** or closely supervise it till it enters another **SSW.**

In achieving cost-effectiveness, don't cut corners by accepting myths. Myths which will weaken an ESD-control program are too numerous to be catalogued here, but some of the more dangerous are:

1. "A wrist strap will render the human skin harmless." The truth is that ESD is first and foremost a people problem, and a grounded operator is only the lesser of two evils versus an ungrounded operator because of the **CDM.** Furthermore, the mere wearing of a wrist strap doesn't guarantee that the **ARTG** is being met; continuous monitoring is necessary.
2. "Higher humidity will solve all ESD handling problems." The truth is that increasing the humidity is far from a panacea and will only help. It would be more correct to say that low humidity will aggravate problems when ESD-control measures are marginal.
3. "Components are safe once they are mounted on a printed circuit board." The surprising truth is that they can be at greater risk, because circuit lines increase the target area for **DI** and **CDM** events and act as antennas for **FIM** events. At best the board can provide protective networks which will reduce the ESD hazard but not eliminate it.
4. "Faraday cages are constructed of carbon-loaded plastic." The truth is that carbon-loaded **conductive** plastic provides some **field** attenuation but insufficient protection from **DI.** See **"Faraday Cage"** in Chapter 2.
5. "**Ionization** will allow the use of plain, **nonconductive** plastics in an ESD-controlled area." The truth is that **ionization** will neutralize charges on ungrounded **nonconductors** but will not prevent **triboelectric charging** in the first place. This charging can occur in an ionized atmosphere and zap a part before **ionization** can neutralize the charge! Therefore, **ionization** cannot eliminate requirements for passive control measures.
6. "**Ionization** is a substitute for the use of wrist straps or other **grounding** of personnel." This is an egregious myth that has been proposed in the past by at least one equipment supplier. The truth is that **grounding** is necessary whether the atmosphere is ionized or not, because **ionization** takes time

to neutralize a charge and **ESDS items** can be damaged before the ions have done their work. Remember that the **Basic Rule** requires protection "even for a nanosecond."

7. "Devices with protective networks (designed and built in) are not **ESDS.**" The unfortunate truth is that protective networks make devices less sensitive but can't provide total protection.

There is no one right way to control ESD, and in developing a cost-effective program one must innovate while keeping the ground rules intact. Remember that **operator disciplines** must be compatible with the paraphernalia used in the work place. For example, if smocks are the norm, make them ESD-protective smocks; then personal clothing will be of much less concern and sleeves needn't be short or rolled up (paragraph 3.6.12.10 of Model Specification 1 in Chapter 6). If reusable packaging is employed, the procedure for handling it will differ from that for throwaway materials. Using **conductive** flooring with **conductive** footwear to ground the skin, as opposed to wrist straps, will modify disciplines; for example, shoe straps or soles will have to be kept clean to maintain conductivity of their sufaces. The near-term cost compared to the long-term cost of ownership of equipment must be evaluated up front, to avoid later changes which might impact productivity. For example, nuclear ionizers have an annual cost for the polonium-210 elements. If this cost is discovered after a few years and electrical ionizers substituted, rules must be imposed for avoiding **FIM** damage from **fields** created by these new ionizers, and failure to add these rules could cause loss of **ESDS** hardware (see paragraph 3.6.8 in Model Specification 1 in Chapter 6). Control measures must be tailored to the work area and keep up with changes in it.

Plan of Action

A plan of action can now be formulated to assign the responsibilities and actions required to put the ESD-control program into place. This plan should include, but not be limited to, the following elements:

1. Proposals. ESD-control requirements must be considered when a program is still in the proposal phase. Expected levels of sensitivity should be identified throughout the in-house production process as well as repair and maintenance activities. Failure to identify these requirements in the proposal phase will cause unexpected costs to occur throughout the life of the program and may have a negative impact on the company's reputation for years to come.

2. Contracting. The negotiators for both the contractor and the customer must have a meeting of the minds on ESD control before the contract is signed. An imprudent contractor can give away the store with overkill that imposes unnecessary, excessive costs on himself or he can underkill and leave himself open to quality and reliability problems. Contract boiler plate drafted in the vacuum-tube era is no longer appropriate.

3. Program Management. Program managers can make or break an ESD-control program. A realistic, cost-effective program will be accepted and supported only if it is well-defined and properly communicated to program management. Involving program managers in the planning and development phase of the program will expedite its future implementation in all functional areas.

4. Purchasing. Buyers of ESD-protective materials and equipment must have a degree of expertise in every facet of the ESD control program for the purpose of estimating the cost of coordinating the requirements with suppliers and subcontractors. The supplier and subcontractor ESD-control program is generally the responsibility of the prime contractor by Department of Defense mandate. The purchasing function should be established as the key interface between the supplier and subcontractor ESD-control program and the program of the prime contractor.

5. Logistics. All documentation, including technical publications such as service and repair manuals, and lists of spare parts, must be in agreement with the ESD-control program. The field service organization must train its engineers and technicians in a wide variety of ESD-control methods. Field environments can be radically different from that of a factory and may require a much higher level of expertise.

6. Facilities and Industrial Engineering (F&IE). Personnel in F&IE will be called upon to provide facilities and equipment to establish and maintain ESD-controlled areas. The ability to debunk the myths about **humidification, ionization,** and the like, is a must for all F&IE personnel. Facilitization of an ESD-protected area can be one of the more costly elements of the program and therefore should be specified and designed by people well-trained in electrostatics.

7. Plant Services. Maintenance personnel will be required to service and repair some rather unusual equipment. Electricians must understand the special **grounding** requirements of ESD control and the need to suppress

fields around some motors, transformers, and switching equipment. Janitors need to know about application of **static-limiting floor finish,** the hazards of **nonconductive** trash bags, and so forth. All personnel working in ESD-controlled areas must be aware of the threat of ESD to company products and profits.

8. Engineering. Engineering will determine and specify the *V*-zap levels (see "**HBM**" in Chapter 2) of the product in every configuration from component through module, subsystem, and system levels. Testing procedures, failure analysis, and designing to meet customer ESD requirements are only the tip of the iceberg. The engineer who takes the easy way out when specifying ESD characteristics can send production costs out of sight. Few engineers have had formal training in electrostatics, and even fewer have experienced the unpredictable behavior of electrostatic phenomena, e.g., **tribolelectic charging,** at assembly stations. Therefore, educating engineers in ESD is an absolute must. Even then, standards, handbooks, and specifications will not be understood in the same way by all engineers, and the burden of specification interpretation will fall on the shoulders of Engineering management.

9. Manufacturing. Manufacturing management must make a total commitment to the ESD-control program. A basic system of ESD controls in the factory is required in the absence of detailed drawings or specification call-outs and will become a way of life in the shops. Since ESD-control rules can be circumvented with no tangible evidence and possibly with no harm done, the effectiveness of the program depends on the integrity of the workers. This unique characteristic of ESD control makes program maintenance of the utmost importance, and this maintenance in the factory might be described as a continuous selling job.

10. Quality Assurance (QA). QA, with its many subset activities, will interface with every part of the ESD control program from procurement to final test and customer acceptance. The ESD training of inspectors and first-line management should be more extensive than that of manufacturing operators. As in manufacturing, the integrity of personnel is primary to success of ESD control.

Program development should be a team effort of experts representing each of the ten functions noted above. The plan of action should deal not only with implementation but with long-term maintenance of the program. The electronics industry is changing daily, and the ESD-control program must change with it. The need for minor changes will be obvious only to persons close to the program, and they must ensure that it grows as required to maintain high yields of product and prevent **latent failures.**

In conclusion, an ESD-control program is unlike anything else experienced in the electronics industry because of the **latent failure** aspect: the product cannot be routinely inspected to determine if the program is effective. Thus the program depends on good practice, not testing, and requires integrity by everyone from top to bottom. It is expensive and not easy to justify. If the program is realistic and clearly presented to the people who can make or break it, it will be successful. Also, it must be cost-effective, which is the key feature of the approach we are outlining in this book. Unnecessary overprotection or overkill tends to result in an unprofitable project—but remember that inexpensive overkill, as with foil laminate bags as **Faraday cages,** is all to the good. No matter how effective in theory, the program will receive multilateral support only if it is comfortable to implement. That is, it must be rational, understandable, and practical, and this is the kind of program we are attempting to sketch for you.

The Enforcement Problem

ESD control is simple in principle—just follow the **Basic Rule** and some corollaries such as our **CD Rule**—but a program requires endless judgments in practice. Some of these concern enforcement. Can rules ever be waived? If so, when?

For example, can the **Faraday cage** definition ever be violated? Consider a metal-cased electronic box which is itself a **Faraday cage** except for the connectors (Example 30 in Chapter 5). Should the connectors be covered with metal caps or with aluminum foil to complete the **Faraday cage** as "electrically continuous" by our definition? Or should the box without connector caps be packaged in a **Faraday cage**? By the absolute or "Ten-Commandments" approach, such a package would be foil or better or at least meet the **discharge test** of paragraph 4.3.3 of Model Specification 2 in Chapter 6. However, in Example 30 of Chapter 5 we applied the "situation-ethics" approach in the form of a **special test** which wasn't quite the specified **discharge test.** Thus we bent the rules and allowed a see-through metallized bag as packaging. Incidentally, another **special test** might have involved measuring the attenuation of a worst-case expected **field** by a metal plate with a hole representing the connector opening beneath which the pins are set; high attenuation would justify low **shielding** by the package. The point is that judicious rule-bending is allowable and even necessary for a cost-effective program. In the above example we were avoiding the expense of opening and closing an opaque package to verify the code numbers on the box.

But what about the enforcement of rules such as **operator disciplines** that may not be safely bent? In a milieu where engineers and even assemblers are individualistic, short cuts will be taken without consulting authority. Besides willfulness, the reason for such disobedience is twofold: (1) the need for the rule is not understood and (2) compliance is onerous. Therefore, part of the remedy for noncompliance is thorough training of all personnel so that they grasp the concepts underlying all rules. Another part of the remedy is to design rules which are comfortable, and operators' opinions must be considered seriously in this design process. For example, metal expansion bands for wrist straps serve their purpose of providing good skin contact only if they are worn correctly. In practice, our operators complained that these bands pinched their skin and pulled hairs, so some individuals stretched their bands to deform them until they hung loose, used oversized bands, or, in extreme cases, worked without a wrist strap when the coast was clear. (This was in the days before **continuous wrist-strap monitors.**) We should have evaluated various band designs in limited numbers and then ordered the Velcro-fastening cloth or adjustable stretch-cloth bands which our operators preferred and would have worn willingly.

In conclusion, bend rules only when there is strong incentive as well as justification by **analysis** or **special test.** Get workers' opinions, and if a rule is considered impractical try to replace it. If a rule is troublesome to follow but there is no alternative, explain its purpose to all personnel involved and insist that they comply with it unless they can suggest an alternative. An assembler or technician who is taught the reasons for disciplines and is allowed to help select or change them is likely to be cooperative. As in politics, autocracy breeds rebellion. What is wanted is a measure of democracy but not so much that the management chain of command is disrupted. Also, ESD control should be included among company suggestion programs giving certificates and awards.

Advice to Small Companies

Small companies with limited ESD-control budgets don't have enough money to buy their way out of ESD problems, but that approach fails anyway. When funds are low, our motto "Buy the basics, forget the frills" is especially apropos. Here's an example of two shopping lists for equipment and materials for simple assembly work where **triboelectric charging,** as caused by processes such as grit-blasting, doesn't necessitate the use of **ionization:**

List A

NEEDED FOR EACH OPERATOR	COST, $ (ILLUSTRATIVE ONLY)	
Conductive floor mat	25	
Conductive footwear	25	
Bench-model air-ionizing blower	400	
Conductive chair	100	
Antistatic smock	50	(including cleaning cost for a year)
Ordinary one-conductor wrist strap	20	
Total:	620	

List B

NEEDED FOR EACH OPERATOR	COST, $ (ILLUSTRATIVE ONLY)	
Static-limiting floor finish	20	(excess in cost over standard finish for a year)
Continuous wrist-strap monitor unit, including a special strap and two-conductor cord)	100	
Topical antistat	10	
Field meter	300	
Zapflash	25	
Total	455	

What does list A buy in ESD safety over list B? List A includes a **conductive** mat and footwear as a backup to the unmonitored wrist strap, but dirt on the mat and/or shoe soles can cause resistance to ground to exceed the **ARTG,** and the operator might lift both feet and be completely ungrounded. Use of a **continuous wrist-strap monitor** is the only positive method of enforcing the **Basic Rule,** which does not allow the **ARTG** to be exceeded "even for a nanosecond."

The personal **field meter** puts the operator in command of the **FIM** danger and makes an **antistatic** smock and a **conductive** chair unnecessary. Sleeves will be rolled up, and if necessary, as shown by **apparent charge,** the ordinary chair upholstery could be treated with **topical antistat** to meet the **CD Rule.** List B gives the means to meet the requirements of Model Specification 1 in Chapter 6.

Furthermore, list B includes **static-limiting floor finish** as a low-cost, unobtrusive backup or "safety net." ("Do it right, but keep the safety net tight.") For example, if a supervisor with ordinary footwear entered the **SSW** and touched an **ESDS** lead, damage would be far less likely with the special floor finish than with a **conductive** mat, which tends to produce high charges on rubber-soled shoes by **triboelectric charging.**

The ionizer in list A has no clearly defined purpose and is needless in this case; see "**Ionization**" in Chapter 2.

Finally, list B includes a versatile **Zapflash** for checking **grounding** and, among other things, proving that black plastic items are **conductive** (highly loaded with graphitic carbon) and not merely pigmented black (with about 1% of **nonconductive** carbon).

In conclusion, list B is far superior to A but costs 27% less! This is a good example of getting more for your money with basics instead of frills.

The principle of cost-effectiveness illustrated here is built into this entire book, which we have written with the needs of small companies in mind. But remember than an optimum choice of materials and equipment is the minor part of the battle against ESD. The major part is the training and encouraging of operators. For example, that hypothetical $300 for a **field meter** is well spent only if the meter is used properly! See **"Field Meter"** in Chapter 2. Thus, more important than "Buy the basics, forget the frills" is another motto; "Work smart." A small company may have an advantage here, because there are fewer people to be educated and perhaps something of a patriarchal system that encourages personnel to identify themselves with the company and its policies and to be craftsmen rather than bored, alienated hirelings.

Incidentally, the work surface is most economically a **static-dissipative** table mat or, if solvent- and solder-resistance is required, a hard laminate retrofitted to a standard (but grounded) workbench. Much money can be saved by converting existing benches this way rather than buying new ESD-control benches as a big company would tend to do.

A final example of cost-saving on equipment is the use of unglazed cardboard or hard vulcanized fiber tote boxes. Assuming that tote boxes are allowed to go in and out of **SSWs**, the boxes must be **SSPs (Faraday cages)** if unprotected **ESDS items** are placed in them but need be only **antistatic** if the items are safely inside **Faraday-cage** packages such as foil laminate bags. In the latter case, unglazed cardboard (paragraph 3.6.6.1 of Model Specification 1, Chapter 6) or hard vulcanized fiber (Paper No. 7 in the Appendix) are satisfactory. As of early 1989, approximate prices for tote boxes 11 × 17 inches by 6 inches deep in a quantity of 100 were:

MATERIAL	PRICE EACH, $
Corrugated cardboard (**antistatic** by nature)	1.60
Nonconductive high-density polyethylene	4.50
Hard vulcanized fiber (**antistatic** by nature)	5.60
Antistatic polypropylene	5.60
Black **conductive** (carbon-loaded) polyolefin	11.50

The cardboard-foil Corshield box described in Paper No. 7 in the Appendix is potentially very economical but couldn't be included in the above comparisons because it wasn't routinely available in a suitable stock size.

The point is that small companies can cut costs, with no sacrifice in program effectiveness, by the shrewd selection of ESD-control products. The best needn't cost the most.

The Role of the Program Coordinator

In large companies, an ESD program coordinator for each division should be appointed to manage the program and be accountable for compliance with MIL-STD-1686A (a good ruling document even if no contracts are military). The coordinator should:

1. Serve as leader of a working group made up of representatives from each functional organization required by Table 1 of MIL-STD-1686A to participate in ESD control.

2. Conduct monthly meetings of the working group to (1) exchange technical information, (2) present educational/training seminars, and (3) maintain a high degree of ESD awareness.

3. Establish various ESD-awareness projects to motivate workers to be diligent in maintaining disciplines.

4. Respond to all audit deficiency notices on ESD-control matters.

5. Conduct informal surveys to identify potential problem areas and to assist in remedial action.

6. Maintain a list of approved ESD-control materials and equipment (Reference Document 4 in Chapter 6).

7. Evaluate new ESD-control materials and equipment for possible addition to the approved list (using the criteria of Reference Document 5 in Chapter 6).

8. Support new business proposals by providing ESD cost impact data.

9. Support Procurement Quality Assurance (PQA) in the qualification of suppliers and subcontractors to assure comprehensive ESD control from the device level to the completed system.

10. Review company and Government standards and specifications which might impact ESD control.

11. Provide reports and status briefings as required by upper management.

12. Participate in EOS/ESD Association symposia to keep the company program up to date.

13. Review employee suggestions on ESD-related matters. This is important in winning workers' cooperation.

14. Coordinate with the materials and processes laboratory on **special tests** in support of ESD-control disciplines. This includes troubleshooting (Chapter 3).

15. Support the development and revision of operator work instructions. (Work instructions give detailed how-to methods of meeting specification requirements.)

16. Support Quality Assurance (QA) in audits and reviews as required by the company's ESD-control program specifications.

17. Participate in design reviews of **ESDS** equipment as required by customer contracts.

Safety

We mention safety last, but of course it is not least. Be sure to rigidly enforce the safety provisions of Model Specification 1, including use of ground fault circuit interruptors (GFCIs) if necessary, so that personnel cannot be electrocuted by connection to ground as a result of ESD precautions; accidents can happen, of course, whenever a hot lead and a hard ground can be touched simultaneously. Remember to provide springs on the **continuous wrist-strap monitors** for plugging in the wrist-strap cords, as required in Reference Document 3 in Chapter 6, so that the operator can pull free from any direction. Discourage the use of **conductive** materials in favor of **antistatic** or **static-dissipative** ones that cannot carry a lethal current (above 1.0 mA). If **conductive** floors are used—which we don't recommend unless absolutely necessary—beware of low resistance to ground that might allow a person with wet, leather-soled shoes to be electrocuted; we found an example of this condition (Reference Document 1 in Chapter 6).

In conclusion, when you design a safe system consider every possible path to ground and every worst-case scenario, e.g., wet leather-soled shoes on a **conductive** floor as mentioned above. If 1-megohm resistors are installed in every path to ground and GFCIs are used when high voltage is being handled, electrocution as a result of ESD control is virtually impossible—but this is a big "if" that requires vigilance and enforcement. Therefore, QA and the company Safety function must be deeply involved in the ESD-control program.

Finally, we must mention that many marginal safety questions will arise. **Ionization,** for example, is considered somewhat dangerous when polonium-210 is the ion source; see "**Ionization**" in Chapter 2. Many would call this concern far-fetched, and an even more unlikely danger is the low voltage (0.1–5.0 V, Reference Document 3 in Chapter 6) imposed on the wearer of a wrist strap with a **continuous wrist-strap monitor;** in this case the voltage is not only low but it is imposed across two skin-contact points on the wrist so that current will flow only between these and cannot pass through vital organs. Concern over such matters may seem ridiculous, but some non-engineers are easily alarmed by any new "hazard" in the work place, and Safety must take their concerns seriously and provide reassurance. Genuine hazards, such as those noted in GIDEP Alerts, must be promptly removed.

Chapter 8

Disposition of Mishandled Hardware

This is a vital subject when expensive products, e.g., modules worth $100,000 each, might have to be scrapped because of ESD damage that is measured or even only suspected.

The general definition of "mishandling" includes overstress by dropping, bumping, temperature excursions (heating or cooling), vibration, or other mechanical or environmental events, but we're concerned here only with ESD.

It is assumed that the attending employee, or another employee who observes the ESD overstress, will take personal responsibility and promptly report the incident to a Quality Assurance (QA) representative, who will fill out the appropriate paperwork ("squawk sheet"). Then a Reliability Engineer will review the data and make a judgment. These data are:

1. Identification including serial number of mishandled **ESDS item.**
2. Violated specification requirement(s) with paragraph number(s).
3. Date and time of day of the infraction.
4. Relative humidity and temperature.
5. Nature and arrangement of nearby objects during the infraction, including distances of charged surfaces from the mishandled item.
6. **Apparent charges** on these charged surfaces measured as soon as possible after the infraction.
7. **Surface resistivity** of workbench top, floor, and surfaces of nearby objects.
8. In a reenactment of the infraction, **apparent charges** on objects and voltage on the operator.
9. Identification of parts of the **ESDS item,** e.g., terminals or circuit lines, which were touched by the operator during the infraction.
10. Functional test data on the item after (and, if possible, before) mishandling.

These are the basic data to be recorded in the squawk. They may or may not be sufficient for Reliability, with advice from ESD specialists in other departments, to make a decision on disposition of the item.

In the case of the module in Example 25 of Chapter 5, functional test showed that all 83 CMOS devices on the board had been damaged by electrical overstress because of the **field** generated by grit-blasting. These components made up 51% of all component leads, so rework was judged impractical and the module was scrapped. This is an example of **analysis;** no testing beyond the functional test was required.

On the other hand, all components of the mishandled **ESDS item** might pass the functional test, or a limited number might fail so that rework could be feasible. Now the question is: might those components which passed the functional test have been subtly damaged by the ESD event so that **latent failure** is possible? The next step is to run more thorough tests, e.g., with a curve-tracer, to look for damage not seen in the functional test. Probably those components showing even subtle damage should be replaced, but the decision will depend on reliability level of the product, rework or replacement cost, etc. Finally, the most difficult question must be asked: though parts were not damaged measurably, even by thorough tests, might they still have been subtly "wounded" by ESD? A judgment that they were not so damaged might be based on **analysis.** For example, if the measured voltage on the operator and the **apparent charges** on all charged objects in the vicinity were less than 300 V, and the **ESDS item** was sensitive to 500 V, the item could be assumed to be undamaged. But suppose the voltage on the operator reached 2000 V, as measured by a **Personnel Voltage Tester** Model PVT-300 by Voyager (Paper No. 10 in the Appendix), and he had touched a sensitive lead. Then the item might have to be scrapped even though no damage had been found by curve-tracer or any other test method. Of course, considerations such as reliability level of the item would enter the decision as mentioned above.

On the other hand, suppose that the operator was grounded but accidentally dropped a film of common **nonconductive** polyethylene at 2000 V **apparent charge** onto the 500-volt-sensitive item. Now the picture in terms of the 2000 V is unclear because charges on **nonconductors** are indirectly and incompletely manifested on **ESDS items** via **fields** (the **FIM**). In this case, a **special test,** specifically a test using a **coupon** like those in Paper No. 8 in the Appendix, would be useful. No damage to five **MOSFETs** in a row by dropping similar polyethylene film at 2000 V **apparent charge** onto the **coupon** would give confidence that the real **ESDS item** hadn't been subtly damaged. The necessary assumption is, of course, that lack of *measurable* damage to the very sensitive **MOSFETs** means lack of *unmeasurable* damage, which could lead to **latent failure,** to the less-sensitive **ESDS item;** the item in this example is sensitive to 500 V, whereas the **MOSFET** is sensitive to 100 V and its sensitivity is heightened still more by the large antennas and less **voltage suppression** of the **coupon.**

A background which is relatively benign in terms of ESD supports arguments to save mishandled hardware. Therefore, safety nets such as **humidification, ionization,** or **static-limiting floor finish** are desirable. But remember that **humidification** and especially **ionization** must be used with caution; see the discussion of these subjects in Chapter 2. For example, the reenactment of an infraction in which the operator touching **ESDS** leads was accidentally ungrounded might show 100 V on his skin with **static-limiting floor finish** versus 800 V with standard floor finish. Then, if the mishandled item were sensitive to 500 V, damage by **DI** from the operator's fingers could be discounted for the **static-limiting floor finish** but not the standard one; the safety net proved useful.

Obviously, judgments often will have to be based on incomplete data, just as business decisions are made, but safety nets and worst-case **special tests,** e.g., with **coupons** as described above, will help greatly. The more critical and high-reliability the **ESDS item,** the less leniently should infractions be treated. Remember that **latent failures** are real though they may be rare (see **"Latent Failure"** in Chapter 2). Ethics forbid the use of sophistry to excuse any and every infraction not causing measurable damage, and some mishandled hardware will have to be scrapped even though it passes all its functional tests.

Chapter 9

Check Lists, What to Buy and Do, Conclusion

Check Lists, General

Failure Mechanisms (DoD-HDBK-263)

Voltage-dependent:
- dielectric breakdown
- gaseous arc discharge
- surface breakdown

Power-dependent:
- thermal secondary breakdown
- metallization melt
- bulk breakdown

Damage Mechanisms (Chapter 2)

Discharge type:
- **DI** (imperfectly modeled by **HBM**)
- **CDM** (involves **DI**)

Field type:
- **FIM**
- **FFB**

Greatest ESD Hazards

Personnel (**DI** by touching)
CDM especially by leads touching **conductors**
Fields from dynamic processes such as grit-blasting or spray-coating **(FIM)**
Fields from charged **nonconductors** such as common plastics **(FIM)**
Fields from packing material such as polystyrene foam ''peanuts'' **(FIM)**

Key Elements of Our Approach

Basic Rule
SSZ concept
CD Rule
Safety nets such as **static-limiting floor finish**
Operator disciplines
Troubleshooting with **analysis** or worst-case **special tests**
Cost-effective materials and equipment
Techniques such as **voltage suppression**
Conservative **ARTG** (10 megohms, maximum)
Frequent use of **field meter, Zapflash, topical antistat**

Minimal *SSW (Chapters 4 and 6)*

Static-dissipative table mat
Lead wire with resistor
Distribution wire with clamp to ground
Field meter
Continuous wrist-strap monitor with strap (cord and band)
Topical antistat
Static-limiting floor finish
Zapflash
Well-trained operator!

SSP Requirements (Chapters 3 and 6)

Adequate **shielding,** preferably by **Faraday cage**
Intimate wrap preferably **antistatic**
Secure mounting of **ESDS item** in package
Cushioning by **antistatic** bubble-wrap or resilient foam
Compliance with **triboelectric charging** test (if necessary)
Compliance with **special test** with **MOSFETs** (for highest confidence)
Freedom from contamination or corrosion of packaged item

Documents for Reference

MIL-STD-1686A
DoD-HDBK-263
MIL-HDBK-773
ESD Controls Study, Final Report, prepared for NASA by Reliability Analysis Center, Rome Air Development Center, September 1981

NAVSEA SE 003-AA-TRN-010: Electrostatic Discharge Training Manual
MIL-STD-129: Marking for Shipment and Storage
EIA 541: Packaging Material Standards for ESD-Sensitive Items (was Interim Standard 5-A or 5-A-1) (order from Electronics Industries Association, Washington, DC)
NFPA 99: Standard for Health Care Facilities (order from National Fire Protection Association, Quincy, MA) (formerly NFPA 56A; Chapter 12 defines electrode for measuring resistance of surfaces to ground)
DoD-STD-2000-1B (soldering) (replaced by MIL-STD-2000 in January 1989)
MIL-M-38510 (microcircuits)
MIL-B-81705 (packaging)
FED-STD-101 (test procedures)
(Other Government specifications and standards as required)

Check Lists, What to Buy

Materials and Equipment

Workbenches with **antistatic** or **static-dissipative** surface and buried **conductive** layer (for economy and portability, use a **static-dissipative** table mat)
Wires and resistors for **grounding**
Continuous wrist-strap monitors with wrist strap (for every operator)
Humidification equipment (if justified)
Ionization equipment (if clearly required)
Antistatic smocks (optional)
"Conductive" shoes (optional)
Faraday-cage bags, boxes, and tote boxes
Conductive chairs (optional)
Static-limiting floor finish
Antistatic bubble-wrap
Topical antistat
Shunting bars or foam
Miscellaneous items (scrub brushes, screens for CRTs, approved desiccator cabinets, etc.)

Test Equipment

Field meter (preferably one for every operator)
Zapflash (one for every operator)

Surface resistivity meter (Voyager SRM-110 or SRM/RTG, which also measures resistance to ground, or equivalent) (one for every ESD-protected area)
Multimeter
Megohmmeter and NFPA 99 electrode for measuring resistance to ground of surfaces (or use Voyager SRM/RTG Meter)
Faraday cup
Personnel Voltage Tester (Voyager PVT-300 or equivalent)
Discharge test apparatus, including high-voltage power supply
MOSFETs and curve-tracer
Charged Plate Monitor (for testing ionizers)
Topaka slip test equipment (for **static-limiting floor finish**)
Gloss meter (for the floor finish)
Samples and ramp for roller test for **triboelectric charging** (see Paper No. 6 in the Appendix)
Miscellaneous equipment for **special tests**

Check Lists, What to Do

Major Tools for Troubleshooting

Analysis
CD Rule
Coupon
Discharge test
Field meter
Grounding
Ionization
Shielding
Special test
Standard test
Topical antistat
Voltage suppression

Operator Disciplines

Maintain **SSZ** and know its boundaries at all times.
Don't touch **ESDS** leads unnecessarily.
Handle parts by cases when size permits.
Don't touch **ESDS items** to unapproved surfaces such as clothing.
Don't fidget or shuffle feet unnecessarily.
Don't touch **ESDS** leads while rising from chair or lifting foot.
Wear wrist strap with **continuous wrist-strap monitor.**

Keep hair less than 2 inches from skin and tie back as required.
Don't remove shorting clips until ready to use item.
Exclude **nonconductors** from **SSW**.
Keep **SSW** clean and orderly.
Roll up sleeves or wear an **antistatic** (ESD-protective) smock.
Follow **CD Rule** using **field meter** to monitor **SSW** for common plastics and other **field** sources.
Ground all **conductors.**
Use **topical antistat** as needed.
Closely supervise transients in the **SSW.**

Use of **Field Meters**

Meter must be calibrated.
Meter must be grounded if it has a ground connection; otherwise, operator must be grounded.
Meter must be zeroed, e.g., on grounded operator's hand.
An ungrounded operator with an ungrounded meter may roughly estimate the voltage on himself by reading a grounded surface.
The **apparent charge** on a surface must be read according to the meter supplier's instructions, e.g., "Multiply scale reading by 2 when sensor is 12 inches from surface being measured."
Readings must not be construed as actual voltage, especially of **conductors.**
Readings must be used only as **apparent charge** in determining safe working distance by the **CD Rule.**

Mottoes

Buy the basics, forget the frills.
Do it right, but keep the safety net tight.
Humidify if you can justify; ionize only when it's wise.
ESD-wax the floor; it doesn't cost much more.
People are the prime ESD problem—and the prime solution.
Work smart.
Don't look for trouble. (Don't touch leads unnecessarily, etc.)
A grounded operator is the lesser of two evils versus an ungrounded operator. (Because of the **CDM.**)

Setting up a Program

Reasons.
Required by Government contract.
Saves money by lowering reject rates.
Prevents **latent failures.**

Overview.

Use realistic, systematic approach.
Gather information:
- *V*-zap levels (by **HBM**).
- Best control methods.
- Needed levels of personnel training.
- Existing policies supportive of program?
- New policies needed.
- Special personnel skills needed.

Arrange information in meaningful format.

What, When, and How.

"What" determined from functional profile.
"When"—start with most-vulnerable areas.
"How"—controversial; buy products to clearly implement the **Basic Rule.**

Factors in Cost-Effectiveness.

Lowest level of ESD-sensitivity.
Mission criticality.
Cost of **ESDS** components.
Cost and feasibility of rework.
Schedule impact of ESD failures.
Willingness and ability of workers.

Assumptions about Our Sample Program.

Lowest level of sensitivity: 100 V.
Mission criticality: high.
Cost of **ESDS items:** moderate to high.
Cost of rework: high.
Feasibility of rework: variable.
Schedule impact of rework: manageable.
Worker skill: good (but may require union cooperation).

Key Operator Disciplines.

Touch **ESDS** devices only when necessary.
Avoid leads and handle by case if possible.
Touch leads only to approved, grounded materials.
Avoid unknown **fields;** use **CD Rule.**
Outside **SSW,** keep **ESDS items** in an **SSP** or closely supervise them.

Dangerous Myths.

"Wrist straps render skin harmless."
"High humidity solves all ESD handling problems."

"Components are safe when mounted on boards."
"Carbon-loaded plastic provides a **Faraday cage.**"
"**Ionization** allows use of plain, **nonconductive** plastics."
"**Ionization** is a substitute for **grounding** (e.g., by wrist straps)."
"Devices with protective networks are not **ESDS.**"

Elements of Plan of Action.

Proposals
Contracting
Program management
Purchasing
Logistics
Facilities and Industrial Engineering (F&IE)
Plant Services
Engineering
Manufacturing
Quality Assurance (QA)

Enforcement Problem.

Bend lesser rules while still meeting **Basic Rule.**
Justify rule-bending by **analysis** or **special test.**
Teach personnel reasons for rules.
Make rules "comfortable."
Get workers' suggestions.
Use democracy but not too much.

Advice to Small Companies.

Buy the basics, forget the frills.
Work smart: train operators well.
Follow our approach.

Role of the Program Coordinator.

Lead working group.
Conduct monthly meetings.
Establish ESD-awareness projects.
Respond to audit deficiency notices.
Conduct informal surveys to find problem areas.
Maintain list of approved materials and equipment.
Evaluate new materials and equipment to add to list.
Support new business proposals with ESD cost impact data.
Support Product Quality Assurance (PQA) in qualification of suppliers and subcontractors.

Review standards and specifications.
Brief upper management on ESD issues.
Attend EOS/ESD Association symposia.
Review employee suggestions.
Coordinate with materials and processes laboratories function on troubleshooting.
Help originate and revise work instructions.
Support QA in audits and reviews.
Participate in design reviews of **ESDS** equipment.

Safety
Use GFCIs when necessary (high voltage).
Install springs on **continuous wrist-strap monitors.**
Discourage **conductive** materials in the **SSW.**
Beware of **conductive** floors.
Consider all possible paths to ground and worst-case scenarios for low-resistance **grounding** of people.
Be sure the Safety function is deeply involved in the ESD-control program.
Deal with employees' concerns about remote safety hazards of new ESD-control equipment.

Conclusion

What are we guarding against? **Fields** and **discharges.** The **SSP** fully protects against both. When out of the **SSP,** an **ESDS item** is either in transit under continuous operator supervision or else is in an **SSW.**

In the **SSW,** where **operator disciplines** are vital, **fields** are controlled by **grounding conductors,** by excluding unnecessary **nonconductors,** and by controlling necessary **nonconductors** with the **CD Rule** and **topical antistat.** The antistat makes **surface resistivity** low enough for incipient **static charges** to bleed off; also, lubricity of the antistat layer on the surface limits the **triboelectric charging** propensity. To control static-generating processes such as grit-blasting, the techniques are **grounding, voltage suppression,** and, if **special tests** demonstrate the need for it, **ionization.**

Discharges from people are avoided by using **continuous wrist-strap monitors** to meet the **ARTG** and keep skin near ground potential along with other grounded surfaces. Also, isolated **conductors** could participate in **discharges** if charged by **fields,** so **fields** from electrical machinery, CRTs, ionizers, etc., must be controlled. Of course, there shouldn't *be* any isolated **conductors,** but human fallibility requires redundant (overlapping) precautions.

Backups to the above primary precautions or defenses include **static-**

limiting floor finish, certain **operator disciplines** such as prohibitions against touching, and possibly **humidification.** We do not recommend **ionization** as a general backup; it should be used only for specific purposes when other methods fail.

Overall, the strategy is to be cost-effective by keeping **fields** weak rather than eliminating them and by substituting astute **operator disciplines** for expensive materials and equipment. In fact, the operator's performance is critical no matter how excellent the facilities.

We've called this book an "approach" and provided ample concepts, techniques, and general advice. The closest we could come to a handbook or set of instructions is the content of Model Specifications 1 and 2 in Chapter 6, but these will have to be much altered by the user. The reason is that setting up an ESD-control program is far from as simple as assembling a piece of lawn furniture (and even that may not be easy). Each program is a special case because of variables such as:

1. Sensitivity classes of **ESDS items.**
2. Reliability level and mission criticality of the items.
3. Commercial versus military requirements.
4. Nature of facilities (available space, ambient humidity, etc.).
5. Unique ESD-control problem situations.
6. Nature of processes: labor-intensive versus automated.
7. Skill level and attitude of personnel.
8. Complexity of the organization (large versus small companies).
9. Extent of management support of program.
10. Funding available for program.

Given these variables, we can't be more explicit than we've been. Besides, no matter how much detail we gave, we could only tell you what to do technically, not contractually or "politically." You must absorb our advice and then cast it into your own mold, resulting in specifications extrapolated from our models.

This book may serve as a "consultant's bible" because an ingenious ESD-control specialist could review a customer's specific needs and adapt our approach to them. In other words, we've provided the cloth to be tailored to fit any program.

In conclusion, our approach is cost-effective yet conservative. By means of strict **operator disciplines** and concepts such as the **SSZ** and the **CD Rule** we preempt purchases of frills such as **conductive** chairs which might otherwise be bought "just in case." But we're conservative in the sense of insisting on protection with high confidence by such means as **Faraday-cage** foil packages (or at least packages passing a **discharge test**), **continuous wrist-**

strap monitors, and a relatively low **ARTG.** We are also wary of dangers such as space charging by **ionization,** corrosion by **humidification,** or contamination by antistats or particles or fibers from **conductive** bags, boxes, wrist bands, etc. "Buy the basics, forget the frills." The money saved on frills such as a room **ionization** system can procure not only essentials, e.g., **continuous wrist-strap monitors,** but highly recommended options such as a one-to-one ratio of **field meters** to operators.

Besides a cost-effective program, we also stress troubleshooting to deal with any situation using the A-to-Z real and conceptual tools in Chapter 2. In summary, what we're propounding is a logical, common-sense, economical approach based on data from meaningful, realistic tests—as opposed to unsupported intuition and data from misleading tests contrived by tendentious suppliers.

We repeat that we are defending against **fields** and **discharges,** and every requirement in the model specifications of Chapter 6 has a purpose in this

Table 9-1. Relation of Defenses to Hazards and Damage Mechanisms.

HAZARD	DAMAGE MECHANISMS	DEFENSES
Fields	**FIM, FFB**	**CD Rule (apparent charge, field meter)** **Ionization** (when clearly needed) **Shielding** **Topical antistat** **Voltage suppression**
Discharges	**DI** (imperfectly modeled by **HBM** for people), **CDM**	Less than 50 V on **conductors** **ARTG** **Continuous wrist-strap monitor** **Grounding** **Shunting** **Zapflash** **Static-limiting floor finish**
Both **fields** and **discharges**	**FIM, FFB, DI, CDM**	**Basic Rule** **SSZ** **SSW** **SSP** **Humidification** (when justified) Approved ESD-control materials and equipment, e.g., **surface resistivity** meter **Operator disciplines**

Table 9–2. Primary and Secondary Defenses.

PRIMARY DEFENSES	SECONDARY DEFENSES
CD Rule	**Humidification** (when justified)
Approved ESD-control materials and equipment	Exclusion of unnecessary **nonconductors** from whole ESD-protected area as well as from **SSWs***
Operator disciplines	Certain **operator disciplines** such as not unnecessarily touching **ESDS** device leads ("Don't look for trouble.")
Ionization (when clearly needed)	**Static-limiting floor finish**
Continuous wrist-strap monitor	
Grounding	

*Not part of our normal procedures but could be done.

"war." By way of a summary, Table 9–1, which is a variation on Table 2–1 in Chapter 2, relates some defenses to hazards and damage mechanisms; Table 9–2 distinguishes between primary and secondary defenses. (*Note:* These tables are simplified for clarity; not all defenses are listed.) Operators as well as engineers should fully understand these two tables. "Do it right, but keep the safety net tight."

A book on industrial accidents makes the important point that the immediate cause, e.g., a spark which ignites a flammable vapor mixed with air, is just the last of a series of causes such as a broken valve letting gas leak, the failure of a detector to note this leaking gas, etc. By the time the final cause occurred, the accident was inevitable. Therefore, the question is not "What was the cause?" but "What can be changed to prevent another similar accident?" Regarding ESD, the immediate cause of damage by **DI** might be a spark from an operator's finger, but the underlying causes could include failure to require **continuous wrist-strap monitors,** inadequate training of personnel, and the use of ordinary floor wax instead of **static-limiting floor finish.** Thus, the chain of causes of ESD damage can lead back through the failure of primary defenses as well as lack of secondary defenses (safety nets) which make the ESD climate more forgiving. When an aerialist falls to his death, the first cause might be a frayed rope on the flying trapeze while the second cause might be a torn or missing safety net.

A capsule summary of this book is:

> This book is not a conventional text but a cost-effective *approach* to ESD control based on extensive data in the authors' ten published papers, reprinted in condensed form in the appendix. Equipment expense is minimized by superior handling techniques which let you "buy the basics, forget the frills." Chapters include real and conceptual tools from A to Z, troubleshooting (special tests), model specifications, program management, and lists of what to buy and do.

Chapter 10

The Future

Changes in ESD-control materials, equipment, and standards as well as in the **ESDS items** themselves are of course inevitable. The following is a quick sketch of some possibilities.

Materials

As we said earlier in this book, the EIA's single classification "dissipative" (at least 10^5 but less than 10^{12} ohms/square) is sensible, but we continue to use the **antistatic** and **static-dissipative** categories in this chapter because they are still "official" (DoD-HDBK-263) and appear in all the old literature on ESD control.

Antistatic packaging materials which depend on antistats that create a surface sweat layer have problems with the fugitivity of the antistat, causing impermanence of the **antistatic** property (Paper No. 2 in the Appendix) and sometimes corrosion and contamination (Paper No. 9 in the Appendix). Another deficiency is the humidity dependence of the sweat layer, which involves moisture attracted from the air. Therefore, hoped-for improvements in **antistatic** or **static-dissipative** materials include permanence and effectiveness at very low relative humidity.

In fact, permanently **antistatic** or **static-dissipative** plastics are on the horizon. For example, "static dissipation originates within the polymer molecule rather than being added" for a new **static-dissipative** ABS alloy (Ref. 10-1). Also, a new bag being test-marketed as of this writing has the unusual properties of true permanence and no antistat rub-off (because it contains no antistat) as well as biodegradability. This material is listed in paragraph 3.3.1.k of Model Specification 2 in Chapter 6. More such products will surely appear.

Conductive plastic materials are filled not only with graphitic carbon but stainless-steel or other **conductive** fibers, and there are proprietary compositions whose **conductive** additives are unrevealed. Many new products can be expected in this area, a prime application being tote boxes. (But note that we discourage **conductive** surfaces in the **SSW;** see Paper No. 7 in the Appendix.)

Coatings are another active field. The **antistatic** or **static-dissipative** ones we have tested either blistered or shifted to **nonconductive** after a 1-day water soak, representing tote boxes being washed or left outdoors in the rain (which has happened!). **Static-dissipative** packaging film with a radiation-crosslinked coating (Ref. 10–2) is noncontaminating and essentially humidity-independent but has the drawback of becoming **nonconductive** after brief exposure to water or various organic solvents; we found this in our work, and the same result was described in a NASA Materials Science Laboratory report issued in April 1988. Attempts are being made by the manufacturer to increase solvent-resistance, and the product may have been improved by the time you read this.

In general, materials of the future will be noncontaminating, permanent by an oven test, e.g., at 160°F, solvent-resistant, and humidity-independent. The user will be able to pick any **surface resistivity** he wants. However, **triboelectric charging** may be a slight problem. The fugitivity of old-fashioned antistats has been beneficial in controlling charging by the lubricity effect, in which a microscopic amount of liquid antistat rubs off onto the other surface and liquid separates from liquid to cause minimal tearing loose of electrons. Without this purposeful contamination, charging of packaged items will tend to be higher. But whether the loss of lubricity for new, permanently **antistatic** packaging materials will make much practical difference remains to be seen. A compromise between contamination and **triboelectric charging** is to minimize the amount of liquid antistat on the surface, as has been done in the new generation of MIL-B-81705, Type II films mentioned in Paper No. 9 in the Appendix.

Equipment

Equipment will evolve; for example, **ionization** will become safer in regard to space charging (causing charging of ungrounded **conductors**) and more fully automatic in operation. Various test equipment will appear as called for in new standards and specifications. We hope that less-expensive versions of popular items such as **field meters** and **surface resistivity** meters will appear on the market so that these essential tools can be more generously distributed to personnel and preferably given to each and every operator.

Standards

New standards are evolving. Unfortunately, some of them may be unrealistic, as discussed for the **shielding/discharge** test for bags in Chapter 3. The astute consumer will take salesmen's claims *cum grano salis* and devise his own **special tests.** Consider the case of **static-limiting floor finish.** If the objective is to limit charges on personnel wearing their customary shoes, then the "bottom-line" test is the **walk test** (Paper No. 6 in the Appendix) or measurements with a **Personnel Voltage Tester** (Paper No. 10 in the Appendix), whereas **surface resistivity** is useful only insofar as it correlates with the voltage on walking people—and this correlation holds only for a given floor finish, not for different finishes being compared. The slip test (Paper No. 6 in the Appendix) is also a critical bottom-line test, though unconnected with ESD, because a slippery floor can bring lawsuits. Let's forget irrelevant tests and concentrate on those which reflect reality.

Among the newer standards are the new version of Test Method 3015.6 in Notice 7 to MIL-STD-883C (February 1988), EIA 541, MIL-STD-1686A (replacing DoD-STD-1686), and MIL-HDBK-773. The last will "change the ways things are done in-plant" (Ref. 10-3); for example, an interesting requirement in Paragraph 5 of this document discourages the use of **conductive** materials because they can accept spark discharges.

ESDS Items

Several years ago there was speculation that protective circuitry might greatly reduce the burden of ESD control by leaving few **ESDS** devices sensitive to less than 2000 V (by the **HBM**) when installed. However, at the same time device geometries were shrinking to less than 0.5 microns, putting some devices into the 12-volt category. On balance, ESD remains a major threat, little alleviated by protective circuitry, and ESD control is difficult indeed for new components sensitive to well under the old **MOSFET** level of 100 V.

For one thing, a wrist strap grounded through 5 megohms will not protect a 12-volt device from an operator stroking a garment sleeve (represented by a plastic film in our test) on a table in such a way that he would generate 3800 V ungrounded; see the curve for Aclar in Figure 1 of Paper No. 10 in the Appendix. To protect such a device, operator movement would have to be restricted and a low **ARTG**, perhaps 1 megohm, required. Also, **antistatic** garments, **conductive** chairs, and **humidification** (if justified) might be needed.

What about **fields**? Making the assumptions of Paper No. 3 in the Appendix, the equation for the **CD Rule** for a device of any **HBM** sensitivity level becomes:

$$d = (100/S)^{0.5}\,(V)^{0.5}/1.8 = (5.556)(V)^{0.5}/(S)^{0.5}$$

where d = minimum safe distance in inches from a charged surface, S = sensitivity of the device in volts, and V = **apparent charge** on the surface in volts. For S = 12 volts,

$$d = (5.556)(V)^{0.5}/(12)^{0.5} = (1.60)(V)^{0.5}.$$

Solutions to the above equation for 12-volt-sensitive devices are:

V, VOLTS	d, INCHES	d, FEET
1000	41	4.2
5000	113	9.4
10,000	160	13
20,000	226	19

Of course, the coefficient, which we calculated to be 1.60, in the above equation would be verified by tests with actual 12-volt-sensitive devices, just as we did with **MOSFETs** in Paper No. 3 in the Appendix.

To protect against **DI,** allowable voltage on **conductors** would have to be about 6 V when handling 12-volt-sensitive devices. This provision would require careful **grounding.**

In general, the **SSW** for highly sensitive items will be analogous to a Class 10 cleanroom: a place where extreme, almost fanatical precautions are taken. However, these precautions can be extrapolated from the approach described in this book. ESD control will still involve "buying the basics, forgetting the frills," but more basics will have to be bought. ESD-protective smocks, for example, may become a "basic" instead of a "frill" because the **fields** on ordinary clothing may not be tolerated for 12-volt-sensitive items as they now are for 100-volt-sensitive items.

It would seem that a challenging future lies ahead for ESD control.

References

2-1. Comments by J. M. Kolyer in "Should Testing be Done Beyond the Lab?" edited by Donald Ford, *EOS/ESD Technology Magazine,* August/September 1987, pages 15 and 22.

2-2. *Electrostatic Discharge (ESD) Protection Test Handbook,* KeyTek Instrument Corp., 1983, pages 10, 12, 13, 16, and 18.

2-3. W. Simmons and P. Adamosky, "Meeting the New 883 Test Method of IC Static Discharge Testing," *Evaluation Engineering Magazine,* July 1988, page 82.

2-4. D. E. Frank, "ESD Considerations for Electronic Manufacturing," presented to American Society of Manufacturing Engineers Westec Conference, Los Angeles, CA, March 21–24, 1983 (Douglas Paper 7324).

2-5. R. Moss, "Exploding the Humidity Half-Truth and Other Dangerous Myths," *EOS/ESD Technology Magazine,* April 1987, page 10.

2-6. P. S. Neelakantaswamy and R. I. Turkman, "ESD Failures of Board-Mounted Devices," *Electronic Packaging and Production Magazine,* February 1987, page 132.

2-7. D. M. Yenni and J. R. Huntsman, "The Deficiencies in Military Specification MIL-B-81705: Considerations and a Simple Model for Static Protection," presented at the Reliability Analysis Center EOS/ESD Symposium, Denver, CO, 1979.

2-8. J. O. Lonborg, "Static Survey Meters," *EOS/ESD Symposium Proceedings,* EOS-5, 1983, page 63.

2-9. S. S. Sullivan and D. D. Underwood, "The Automobile Environment: Its Effects on the Human Body ESD Model," *EOS/ESD Symposium Proceedings,* EOS-7, 1985, page 103.

2-10. D. L. Lin, M. S. Strauss, and T. L. Welsher, "Big Problem Uncovered: Zapper Data Differs," *EOS/ESD Technology Magazine,* August/September 1987, page 9.

2-11. R. E. McAteer, G. H. Lucas, and A. McDonald, "A Pragmatic Approach to ESD Problem Solving in the Manufacturing Environment, a Case History," *EOS/ESD Symposium Proceedings,* EOS-3, 1981, page 34.

2-12. "Nuclear Air Ionizers Recalled by NRC," *Evaluation Engineering Magazine,* March 1988, page 65.

2-13. W. R. Van Pelt, "Polonium-210 Contamination," *Chemical and Engineering News,* April 11, 1988, page 4.

2-14. N. Jonassen, "The Physics of Electrostatics," distributed at the Sixth Annual EOS/ESD Symposium, Philadelphia, PA, 1984.

2-15. G. Baumgartner, "Electrostatic Measurement for Process Control," *EOS/ESD Symposium Proceedings,* EOS-6, 1984, page 25.

2-16. "Latent ESD Failures: a Reality," *Evaluation Engineering Magazine,* April 1982, page 80.

2-17. G. T. Dangelmayer, "ESD—How Often Does It Happen?" *EOS/ESD Symposium Proceedings,* EOS-5, 1983, page 1.
2-18. B. N. Stevens, "Determining the Surface Resistivity of ESD Protective Cellular Packaging Materials," *EOS/ESD Symposium Proceedings,* EOS-8, 1986, page 136.
3-1. GIDEP Alert H6-A-83-02, "Materials, Plastic, Antistatic," December 27, 1983. (Antistats from bags contaminated NASA instrument mirrors.)
3-2. G. C. Holmes, P. J. Huff, and R. L. Johnson, "An Experimental Study of the ESD Screening Effectiveness of Antistatic Bags," *EOS/ESD Symposium Proceedings,* EOS-6, 1984, page 78.
3-3. S. A. Halperin, "Selecting the Proper Protective Bag: Part II," *EOS/ESD Technology Magazine,* October/November 1988, page 15.
3-4. MIL-HDBK-773, "Electrostatic Discharge Protective Packaging," 1 April 1988.
4-1. N. I. Safeer and J. R. Mileham, "A Material Evaluation Program for Decorative Static Control Table Top Laminates," *EOS/ESD Symposium Proceedings,* EOS-6, 1984, page 85.
4-2. J. R. Huntsman and D. M. Yenni, "Charge Drainage vs. Voltage Suppression by Static Control Table Tops," *Evaluation Engineering Magazine,* March 1982.
4-3. E. H. Russell, "Safely Grounding Static-Control Work Surfaces," *EOS/ESD Technology Magazine,* June 1987, page 10.
4-4. R. Kallman, "Comments on Coping," *Evaluation Engineering Magazine,* August 1988, page 108.
5-1. "Arm Hair Pin-Pointed as New Hazard," *Evaluation Engineering Magazine,* May 1984, page 70.
5-2. *Scott's Standard Methods of Chemical Analysis,* 6th Edition, Volume 1, page 334.
5-3. NAV SEA SE 003-AA-TRN-010, "Electrostatic Discharge Training Manual," pages 35 and 36.
5-4. R. D. Anderson, "Alert Error Corrected: How Sweet It Is!" *EOS/ESD Technology Magazine,* October/November 1987, page 8.
6-1. N. B. Fuqua and R. C. Walker, "ESD Controls Study, Final Report," prepared for NASA by the Reliability Analysis Center, Rome Air Development Center, September 1981, page 30.
7-1. K. Jessen and J. Barto, "Static Control Team Concept—Implementation of ESD Protection in Manufacturing," *Evaluation Engineering Magazine,* November/December 1983, page 94.
7-2. J. R. Giuliano, "SD Program Nets Large Financial Gains," *Evaluation Engineering Magazine,* December 1986, page 26.
10-1. "New Plastics Harvest Adds Variety," *Machine Design Magazine,* July 21, 1988, page 12.
10-2. A. H. Keough, "Antistatic Resin Composition," U.S. Patent 4,623,594 (November 18, 1986).
10-3. "MIL-HDBK-773: Sighs of Relief," *EOS/ESD Technology Magazine,* October/November 1988, page 7.

Appendix

Some of the following papers have been condensed to highlight conclusions and save space. Others, such as Number 3, were judged sufficiently important to be reproduced in their entirety.

Paper No. 1 was reprinted with permission of IIT Research Institute/ Reliability Analysis Center. Paper No. 5 was reprinted with permission of *Evaluation Engineering Magazine.* Paper No. 7 was reprinted with permission of *EOS/ESD Technology Magazine.* Papers No. 2, 3, 4, 6, 8, 9, and 10 were reprinted with permission of the EOS/ESD Association.

For a quick overview, leaf through and read the abstracts.

Paper No. 1

Presented at the 3rd Annual Electrical Overstress
Electrostatic Discharge Symposium, Las Vegas, Nevada,
September 22–24, 1981, sponsored by IIT Research Institute. *EOS/ESD Symposium Proceedings,* EOS-3, 1981, page 75.

SELECTION OF PACKAGING MATERIALS FOR ELECTROSTATIC DISCHARGE-SENSITIVE (ESDS) ITEMS

John M. Kolyer and William E. Anderson
Rockwell International Corporation
Autonetics Strategic Systems Division
Electronics Operations
3370 Miraloma Avenue
Anaheim, CA 92803

Note: The following is a condensed version of this paper. See the *Symposium Proceedings* for the complete text.

Abstract

Seven materials were evaluated. Properties such as transparency, puncture resistance, and heat seal strength were measured. Also, ESDS devices in bags were subjected to the external static field of a model "human finger" charged to 25,000 volts as well as to the extreme condition of a high-voltage continuous discharge. The assumption was that if the packaged devices survived ESD overstressing with no measurable damage they would not be subtly damaged under actual handling/transit conditions. Such damage could reduce lifetime and reliability. For the highest confidence in protection from ESD, at an acceptable cost and with military specification coverage, a double-bagging system was adopted: antistatic polyethylene, which is MIL-B-81705, Type II, as the intimate wrap with a foil bag, MIL-B-81705, Type I, as the exterior "Faraday cage." Another foil bag, proposed as MIL-B-81705, Type III, is effective without a Type II inner bag. A brief review is given of the context of static control plant procedures in which the protective bags are used.

Introduction

The selection of packaging materials for ESDS items is difficult because many products are on the market and the claims made by rival manufacturers are often questionable and even contradictory.

The early products included antistatic polyethylene or "pink poly," which is MIL-B-81705, Type II; conductive (carbon-loaded) polyethylene; and a laminate incorporating aluminum foil, which is MIL-B-81705, Type I. Later, variations on "pink poly" and foil bags were developed; these are "blue poly" and a bag proposed as MIL-B-81705, Type III, and so abbreviated in quotes as "Type III." Also, a new concept in "static-protective" bags appeared: polyester film with an exterior coating of vacuum-deposited nickel, protected by a thin lacquer coat, and an antistatic polyethylene lining; this construction has the advantage of partial transparency. The proponents of the latter bag made the industry aware of the need for "Faraday cage" protection from external static fields/discharges and the insufficiency of "pink poly" alone for worst case" handling conditions. An aluminum-coated, partially-transparent bag is now competitive with the original, nickel-coated version. In 1980, Department of Defense documents were issued, DOD-STD-1686 and DOD-HDBK-263, which emphasized that the interior of a bag must be incapable of triboelectric charging, as true of "pink poly," while the exterior should be a conductive "Faraday cage" when external static fields might be encountered.

Our evaluation included all seven materials mentioned above. The objective was high confidence in protection from external fields/discharges, at an acceptable cost, even for items of the greatest ESD sensitivity. The bags were for general use; they were not for clean room applications, handling of open devices, or the packaging of items sensitive to trace contamination. Functional properties such as heat sealability were checked, and packaged MOSFETs were exposed to external static overstresses to provide assurance of freedom from damage by realistic stresses. The chosen material(s), of course, had to be cost-effective.

Experimental Methods

Figures 1 through 4 show some of the experimental methods. Figures 5 and 6 are a typical example of ESD damage.

Materials Tested

The seven materials are described in Table 1. All were received as samples in the form of bags. "Pink poly" is MIL-B-81705, Type II; by "MIL-B-81705" we are referring, throughout this paper, to the current Revision B. Two leading brands of carbon-loaded polyethylene were tested; the volume resistivities, which were calculated from surface resistivity measurements with bar-clampled samples at low voltage, were 235 and 875 ohm-cm. Using

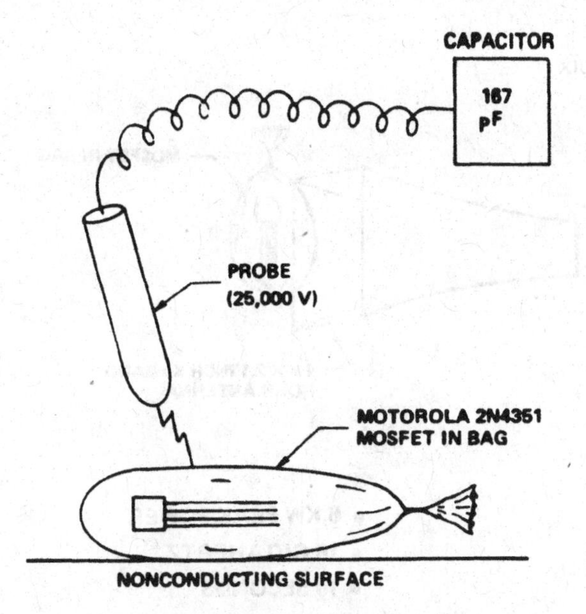

Fig. 1. Charged "finger" test.

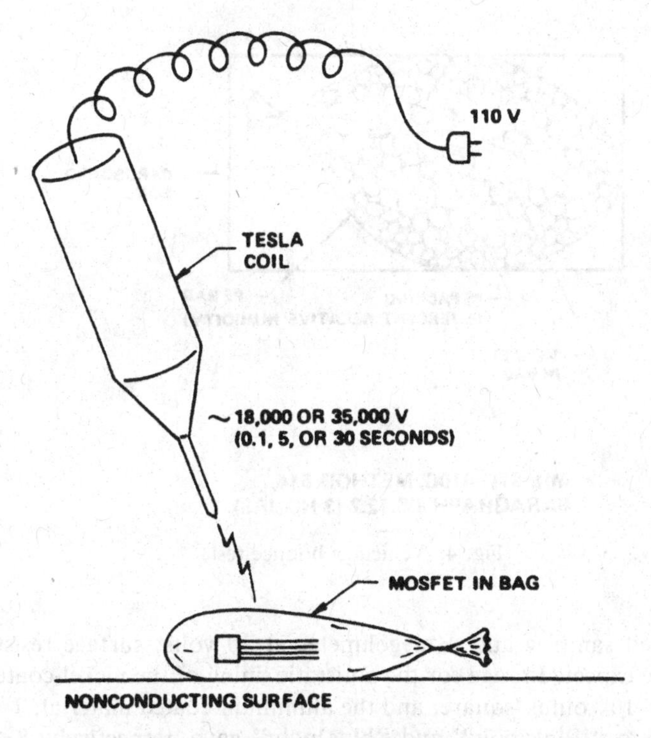

Fig. 2. Tesla coil test.

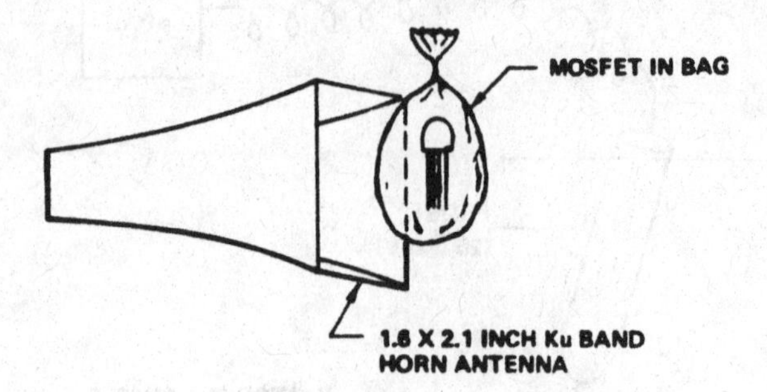

Fig. 3. Radar exposure.

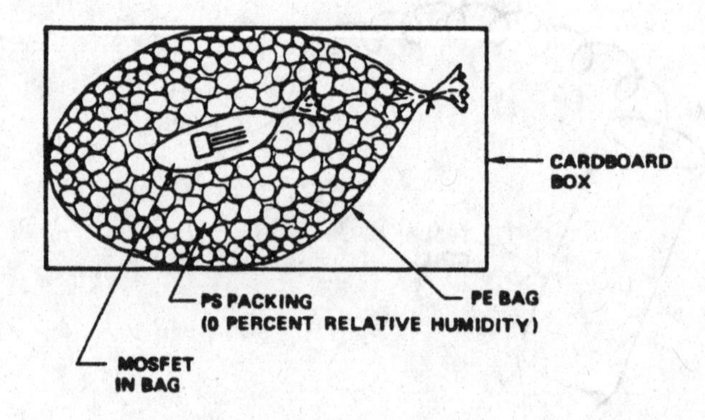

Fig. 4. Vehicular bounce test.

bar-clamped samples and a megohmeter at 10 volts, surface resistivities were in the expected range for the antistatic lining of the nickel-coated material, 2×10^{10} ohms/square, and the aluminum-coated material, 7×10^{11} ohms/square. "Pink poly" and "blue poly" gave, respectively, 8×10^{10} and 1×10^{11} ohms/square. The black lining of the Type I bag gave an

Fig. 5. Punch-through of gate oxide induced by "finger" (25,000 volts, 167 pF) with MOSFET in black bag (1200×).

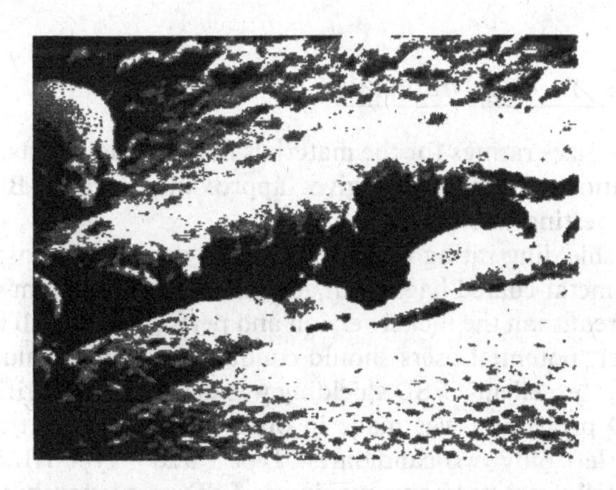

Fig. 6. Punch-through of gate oxide, detail from previous slide (22,000×).

apparent 8×10^7 ohms/square, but edge effects were possible. "Type III" has a "pink poly" lining for which 2×10^{11} ohms/square was found. Surface resistivity measurements are neither closely reproducible nor a critical test for antistatic properties; static bleed-off time (Federal Standard 101, Method 4046) is better for single materials but not for the Type I laminate, whose foil layer drains the charge.

Table 1. Summary of Tests

MATERIAL	ESD SHIELD	TRANS-PARENCY	WATER VAPOR PERM.	ABRASION RESIST.	HEAT SEAL STRENGTH	PUNCTURE RESIST.	ANTI-STATIC PROP.	MIL SPEC	COST*
PINK POLY	P	E	F		E	F	G	YES	13
BLUE POLY	P	G	F		E	F		NO	18
NICKEL-COATED	G	P	F	F	P-E	G		NO	30
ALUMINUM-COATED	G	F	F	F	E	G		NO	28
BLACK	F	X	P		E	F		NO	19
TYPE I	E	X	E	E	G	E		YES	21
"TYPE III"	E	X	E	E	G	E		NO	29

*CENTS PER 8 X 10 IN. BAG, LOTS OF 2000, 5/81

CODE: E = EXCELLENT
G = GOOD
F = FAIR
P = POOR
X = NONE

Conclusions on Bag Materials

Table 1 summarizes ratings for the materials in our various tests, notes military specification coverage, and gives approximate prices. Blank spaces mean that no testing was done.

If an ESD shielding rating of "good" and a degree of transparency are required, the metal-coated bags seem adequate. The aluminum version was more transparent than the nickel version and performed as well in our other tests. However, potential users should conduct their own evaluations.

In our case, "excellent" ESD shielding was desired. No sacrifice in confidence in ESD protection was made in order to gain partial transparency. This position left only two candidates: Type I and "Type III." These foil bags, incidentally, not only give maximum ESD protection but have superior sturdiness (abrasion and puncture resistance) and are excellent moisture barriers. Since Type I has a questionably antistatic lining, as discussed above, only "Type III" with its "pink poly" lining remained as a single-bagging candidate. "Type III" is not yet covered by a military specification, but approval is said to be pending.

Our MX missile module assembly facility, for which a bagging system was being selected, required cushioning material around assembled modules. The choice was "pink poly" bubble-pack with skins of Richmond RCAS-1200 (the only material thoroughly tested by us for antistatic properties). Since this material is not a "Faraday cage," double-bagging was necessary; a Type I bag was chosen as the ESD shield. Type I was preferred to "Type III" because the former has military specification coverage and the latter's "safe" lining was not required. Also, Type I is less expensive (Table 1).

When cushioning is unnecessary, a MIL-B-81705 system comprises a "pink poly" (Type II) inner bag and a Type I outer bag; both materials are bought from suppliers on QPL-81705-6 (issued December 30, 1980). The inner "pink poly" bag provides high at-work-station transparency; an operator can withdraw the pink bag and easily see its contents. If the latter feature is not considered important, the "Type III" bag is effective but does not yet have military specification coverage.

The above benefits of double-bagging with Type II inside Type I are bought at a cost penalty of 13-21% over single-bagging with metal-coated or "Type III" bags. Using the prices in Table 1, a Type I bag at 21¢ and a Type II bag at 13¢ together cost 34¢, which is 4, 5, or 6¢ more than the nickel-coated, "Type III," or aluminum-coated bags, respectively.

The above selections are based on present knowledge, but there are unanswered questions concerning bag materials. For example, how permanent are their ESD shielding or antistatic properties? Accelerated aging tests combined with real-time exposures should give reassuring, or disturbing, answers. Also, new materials may be expected to appear on the market. The "bag of the future" may have most or all of the following properties: low cost, >75% transparency, high ESD shielding effectiveness (like Type I or "Type III"), permanent antistatic and ESD shielding properties, no dependence on a minimum relative humidity, no propensity to contaminate packaged items, minimum sloughing (with no possible shedding of conductive particles), excellent moisture barrier properties, good heat-sealability, and, last but not least, cost-effectiveness.

Paper No. 2

Presented at the 5th Annual Electrical Overstress/Electrostatic Discharge Symposium, Las Vegas, Nevada, September 27–29, 1983, Sponsored by ITT Research Institute. *EOS/ESD Symposium Proceedings,* EOS-5, 1983, page 87.

PERMANENCE OF THE ANTISTATIC PROPERTY OF COMMERCIAL ANTISTATIC BAGS AND TOTE BOXES

John M. Kolyer and William E. Anderson

Rockwell International Corporation
Autonetics Strategic Systems Division
Electronics Operations
3370 Miraloma Avenue
Anaheim, CA 92803

Note: The following is a condensed version of this paper. See the *Symposium Proceedings* for the complete text.

Abstract

The shelf life of antistatic materials has been questioned. Therefore, accelerated life tests were conducted with several commercial bag materials. One criterion for retention of the antistatic property was ability of the bag lining to triboelectrically charge NEMA FR-4 glass-reinforced epoxy circuit board material. It was found that light rubbing, as occurs on withdrawing boards from bags in in-plant handling, gave the same result—a low and presumably tolerable charge—whether or not antistat was present. Contamination caused higher charging. Thus, shelf life was judged to be unlimited, by this test, if bags are stored closed and clean. Another criterion was the charge on the bag itself. The static field is suppressed by foil or metallization, but surface conductivity is required to control the field for antistatic polyethylene alone (MIL-B-81705, Type II). For a leading brand of the latter, incorporating a low-volatility antistat, the shelf life was estimated as at least 8 years if bags are used at above 20 percent relative humidity. Heavy or repeated rubbing, as may be caused by vibration in shipment, is another matter. Transfer of antistat from bag lining to surface of the repeatedly rubbed item seems necessary to prevent high charging. Again, a shelf life of over 8 years was assigned to the tested brand of MIL-B-81705, Type II, but data were insufficient for a prediction for foil or metallized bags. Antistatic polyethylene tote boxes were also studied. Used, worn boxes can accept a triboelectric charge. Such boxes might be periodically treated with topical antistat solution. Conductive boxes are an alternative but have potential problems such as sloughing of conductive particles or triboelectric charging of ESD-sensitive items being added or removed.

Introduction

The permanence of antistatic plastic packaging materials is in doubt because large increases in surface resitivity during a 1-year shelf life of films or foam sheets have been reported.[1] Increased susceptibility to triboelectric charging by polystyrene foam also was found for aged antistatic materials. The indication is that the antistat on the surface of the plastic volatizes into the air or is removed by contact with absorbent materials such as cardboard. Therefore, minimization of exposure to moving air or paper products was recommended.[1]

In view of the above, our purpose was to establish the permanence of certain antistatic materials, in the form of bags, which we use or might use. Of particular interest was the propensity of bag materials to triboelectrically charge a printed circuit board module on insertion or withdrawal. Therefore, a static charging test was devised in which a bare glass-reinforced epoxy board was pulled in a reproducible manner from aged bags and the charges on the board measured. Also, surface resistivities of aged bags were noted. Accelerated aging was achieved by exposing bags, both closed and propped open, to 160°F in a forced-draft oven. Bags were also exposed to blotting paper.

Tote boxes were also of concern because used, worn, antistatic polyethylene boxes found in the plant were observed to take charges of up to a few kilovolts when stroked with a nylon cloth. Whether a given charge was really a practical danger was judged by dropping highly ESD-sensitive devices [metal oxide semiconductor field effect transistors (MOSFETs)] into a non-antistatic tote box charged to various levels and electrically testing the devices for degradation.

Bag Materials Tested

Nine bag materials were tested (Table 1). Five of these had been characterized by use 2 years earlier.[2] All samples except the foam were received in the form of approximately 8 × 10 in. bags; the foam was stapled into pouches. The abbreviations in Table 1, eg, "black" or "grid," are for convenience.

The "Type I" material previously characterized[2] was the Brand B material in Table 1. "Pink poly" was MIL-B-81705, Type II, Qualified Products List (QPL) material, specifically RCAS-1200 from Richmond Corporation.

Table 1.

Type I (brands A and B)	10	MIL-B-81705, Type I (vapor-deposited aluminum + spun-bonded polyethylene + aluminum foil + carbon-loaded polyethylene)
Pink Poly	6	MIL-B-81705, Type II (antistatic polyethylene)
Foil + Antistatic Poly	8	Antistatic spun-bonded polyethylene + aluminum foil + antistatic polyethylene
Nickel-Coated	2.5	100 Angstroms nickel + 1 mil polyester + 1.5 mils antistatic polyethylene
Stainless-Steel-Coated	3	Polyester + stainless steel + antistatic polyethylene
Black	4	Carbon-loaded polyethylene (conductive)
Grid	4	Antistatic nylon + barrier film + conductive ink grid + barrier film + antistatic copolymer
Antistatic Foam	250	Layered, antistat-treated polypropylene foam

The "black" material was the brand for which a volume resistivity of 235 ohm-cm had been found.[2] This bag, of course, is conductive rather than antistatic.

Experimental Methods

Figures 1 and 2 show the principal experimental methods.

Conclusions

Shelf Life of Antistatic Polyethylene Bags

1. In terms of charging a popular circuit board material, under the specific conditions of our test, shelf life is unlimited it the bags are kept closed and clean. Antistatic polyethylene has a long record of successful use. Also, our tote box tests suggest that limited static fields may be tolerable. Therefore, it is presumed that the relatively low triboelectric charges detected on the circuit board would not endanger attached devices. In contrast, the high charges observed when the board was stroked with carbon-loaded polyethylene might well be damaging. However, voltage suppression would limit the field on a multilayer board with internal planes of copper.

2. In terms of surface resistivity, shelf life at usual relative humidities (over 20 percent) is conservatively estimated as 8 years for 6-mil "pink poly" bags when kept closed. This relatively good permanence is probably due to the low vapor pressure of the antistat, which is extruded into the

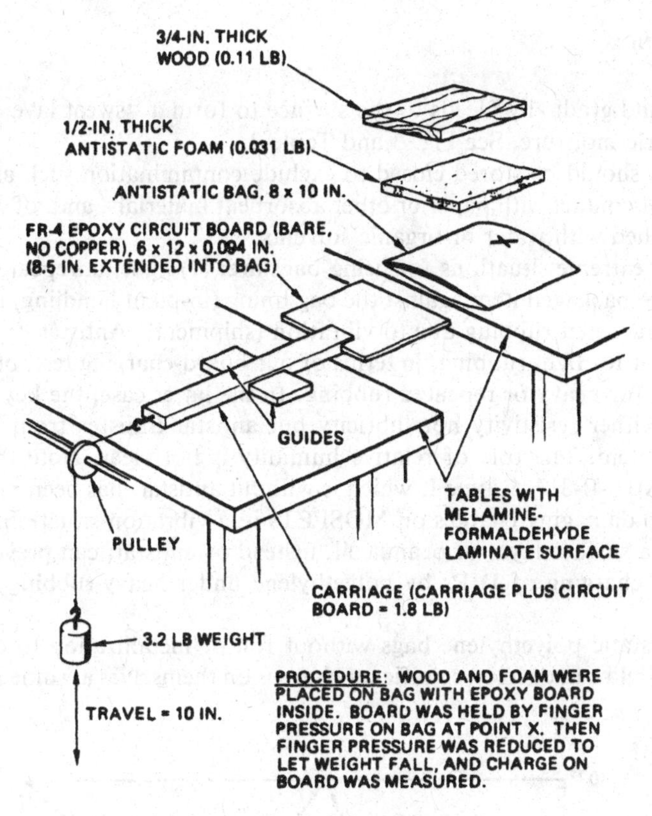

Fig. 1. Static charging apparatus.

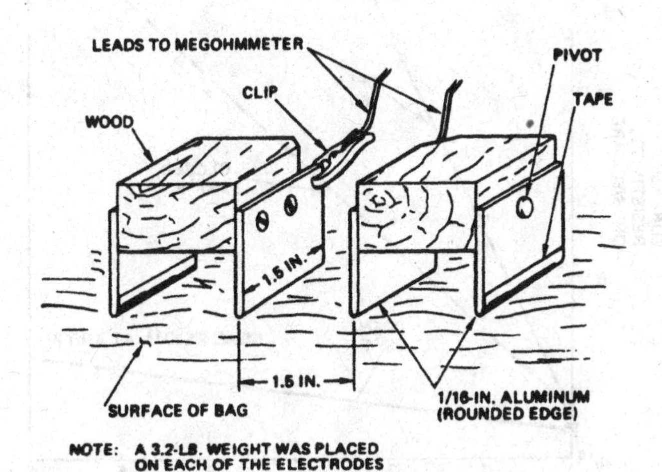

Fig. 2. Surface resistivity electrodes.

polymer and gradually bleeds to the surface to form a "sweat layer" with atmospheric moisture. See Fig. 3 and Table 2.

3. Bags should be stored closed to exclude contamination such as dust, kept out of contact with paper or other absorbent materials, and, of course, never washed with water or organic solvents.

4. Two extreme situations for using bags are: (1) light, unrepeated rubbing by the packaged item against the bag lining (in-plant handling) and (2) heavy or repeated rubbing due to vibration (shipment). Antistat depletion is irrelevant for light rubbing, in terms of our board-charging test, but may be critical for heavy or repeated rubbing. In the latter case, the key factor may be neither resistivity nor lubricity but antistat transfer from bag to packaged item. The role of relative humidity is not clear. Note that the lining of MIL-B-81705, type I, which is without antistat, has been reported to develop damaging charges on MOSFETs in a vibration situation.[3] Also note that a surface layer of peanut oil, instead of antistat, can prevent triboelectric charging of DIPs by polyethylene under heavy-rubbing conditions.[4]

5. Antistatic polyethylene bags without foil or metallization to control the static field can develop significant charges on themselves at some combi-

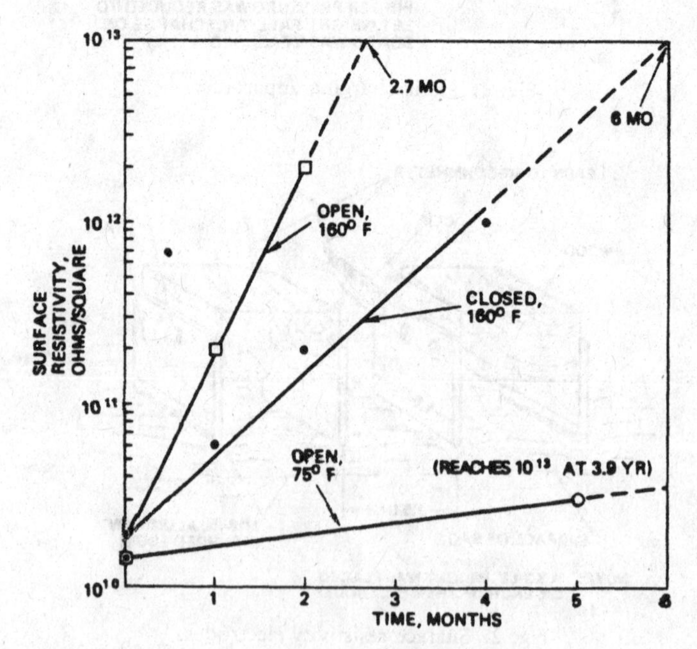

Fig. 3. Surface resistivity versus time for aged 6-mil "pink poly" bags.

Table 2. Estimated Shelf Lives for Antistatic Bags (Used at >20 Percent Relative Humidity).

EXTREME USE CONDITION	ESTIMATED SHELF LIFE (CLOSED, CLEAN)	
	MIL-B-81705 TYPE II	FOIL OR METALLIZED
Light rubbing of FR-4 epoxy/glass on insertion/withdrawal (in-plant use)	8 years	Unlimited
Heavy/repeated rubbing, eg, due to vibration (shipment)	8 years	Not estimated (rate and effect of antistat depletion are unknown)

nation of antistat depletion and low moisture content of the "sweat layer." Here, relative humidity could be a critical factor. Thus our prediction of a long shelf life for "pink poly" (MIL-B-81705, Type II) is limited to use at > 20 percent relative humidity.

ESD Hazard Caused by Static Charges on Tote Boxes

1. Used, worn antistatic polyethylene tote boxes can develop static charges of a few kilovolts when stroked with a nylon cloth. The MOSFET tests indicated that up to 4 kv could be tolerated, but weaker fields could cause ESD damage when "antennas," e.g., circuit lines, are present (see Paper No. 3).
2. It is recommended that antistatic tote boxes be treated periodically with topical antistat solution. This is an inexpensive procedure in terms of antistat cost but may or may not be practical in view of handling and scheduling complexities.
3. Conductive tote boxes have no permanence problem but may triboelectrically charge devices being removed from them, slough conductive particles, or cause ESD damage by rapid discharges as could happen with any conductor.
4. Each user must select a tote box material based on his unique parts-handling situation.

References

1. G. O. Head, "Drastic Losses of Conductivity in Antistatic Plastics," *Reliability Analysis Center EOS/ESD Symposium Proceedings,* pp. 120–123, Orlando, FL, 1982.

2. J. M. Kolyer and W. E. Anderson, "Selection of Packaging Materials for Electrostatic Discharge-Sensitive Items," *Reliability Analysis Center EOS/ESD Symposium Proceedings,* pp. 75–84, Las Vegas, NV, 1981.
3. D. M. Yenni, Jr., and J. R. Huntsman, "The Deficiencies in Military Specification MIL-B-81705: Considerations and a Simple Model for Static Protection," presented at the Reliability Analysis Center EOS/ESD Symposium, Denver, CO, 1979.
4. J. R. Huntsman and D. M. Yenni, Jr., "Test Methods for Static Control Products," *Reliability Analysis Center EOS/ESD Symposium Proceedings,* pp. 94–109, Orlando, FL, 1982.

Paper No. 3

Presented at the 6th Annual Electrical Overstress/Electrostatic Discharge Symposium, Philadelphia, Pennsylvania, October 2–4, 1984, Sponsored by EOS/ESD Association and ITT Research Institute. *EOS/ESD Symposium Proceedings,* EOS-6, 1984, page 7.

HAZARDS OF STATIC CHARGES AND FIELDS AT THE WORK STATION

John M. Kolyer, William E. Anderson, and Donald E. Watson

Rockwell International Corporation
Autonetics Strategic Systems Division
Electronics Operations
3370 Miraloma Avenue
Anaheim, CA 92803

Abstract

Tests were conducted to determine under what practical conditions electrostatic discharge-sensitive (ESDS) items can be damaged at the work station by static charges and fields. Various materials and equipment were evaluated for possible hazards (e.g., static fields created by electrical air ionizers) which must be avoided by careful operating techniques. The data led to important conclusions and guidelines for an ESD control program in accordance with Dod-STD-1686. Examples are given of equipment, materials, and techniques which complement one another, and basic rules for electronic assemblers are suggested.

Introduction

A charged surface always creates a static field (*E*-field). However, in discussing "hazards of charges" we ignore the field and are concerned with discharges between conductors, for example from a person's finger to the copper line on a circuit-board module. In discussing "hazards of fields" we are concerned with ESD damage caused by induction. The effects of RF (radio-frequency radiation) or of magnetic fields (*H*-fields) are not considered in this paper.

Following a description of the test methods used, we evaluate charge and field hazards, including those from electrical equipment, and discuss some methods of coping with them. Finally we briefly describe a synthesis of materials/equipment and operator techniques for handling ESDS items in accordance with DoD-STD-1686.

Test Methods

Static Meter Measurements

Measurements were made with a Simco Electrostatic Locator, Type SS-2 (Simco Co., Inc.), calibrated with a 12-inch-square aluminum sheet charged to 1,000–10,000 volts by a current-limiting high-voltage dc power supply (TARI Central Control Module, Static Control Services). The charges reported in this paper are apparent charges; for example, a 4-inch-square aluminum sheet charged to 10,000 volts showed an apparent charge of about 5,000 volts. The apparent charge is a measure of the strength of the field, which can damage ESDS items by induction.

Surface Resistivity

The electrodes have been illustrated.[1] A 5-lb weight rested on each electrode, and strips of soft carbon-filled polyolefin were placed under the blades to improve contact with static-dissipative or antistatic surfaces. With this arrangement, a steel surface read 10^5 ohms/square, so 10^5 was subtracted as a correction factor. Readings were made with a Beckman Model L-10 megohmmeter at 500 volts unless otherwise noted. Multimeters were used for lower voltages.

MOSFET Damage Tests

The *N*-channel metal oxide semiconductor field effect transistor (MOSFET) used in our tests was a Motorola 2N4351 with the shorting bar removed. The metal TO-72 type package had four leads (source, gate, drain, and substrate-case). This device contains no input/output protective diodes or circuitry. Using a Tektronix 576 curve tracer, gate-source threshold voltages ($V_{GS(TH)}$) were read at V_{DS} = 10 volts and I_D = 10 μA in accordance with MIL-STD-750B, Method 3403. The operator was, of course, grounded with a wrist strap when making measurements.

For many experiments a "MOSFET board" was used; see Fig. 1. The substrate-case lead of a MOSFET was clipped to the "antenna" (circuit line) touched by the operator, while the gate lead was clipped to the circuit line with the projecting lead. The choice of substrate-case and gate leads, rather than another combination of leads, maximized the oxide layer area exposed to an ESD. Each circuit line was 4 inches long, and the lead projected 1.7 inches beyond the end of the board. The circuit-line antennas increased the likelihood of ESD damage to the MOSFETs by static fields.[2] Similar printed circuit boards with MOSFETs and circuit lines for antennas

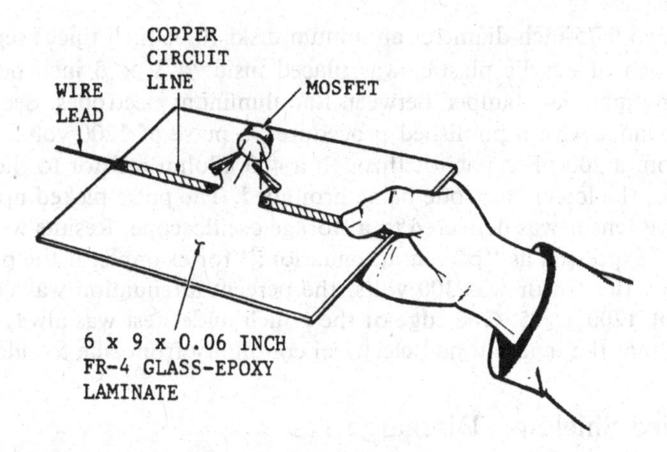

Fig. 1. MOSFET board.

have been used in other investigations[3,4] but for different purposes. Our board, held by the operator as shown in Fig. 1, was intended as a worst-case simulation of ESDS modules being handled at the work station.

MOSFETs were often shorted, as noted in the data, but they were also considered damaged if the current-voltage curve changed and/or $V_{GS(TH)}$ shifted by more than 0.1 volt ($V_{GS(TH)}$ readings for undamaged MOSFETs were reproducible to ± 0.02 volt). Since latent ESD failures are a reality,[5] a $V_{GS(TH)}$ shift of only 0.1 volt is considered significant. Damage is reported as a fraction; for example, "2/5" means that two MOSFETs were damaged out of five tested.

Room Ionization System

The power supply was mentioned above. Two emitters (Static Control Services), each with four corona-discharge points and a set of plastic reflectors, were positioned 3 ft apart on the ceiling of an open-fronted booth 95 inches high, 71 inches wide, and 47 inches deep. This booth contained a bench 29 inches high, 60 inches wide, and 30 inches deep. The distance from the emitter points to the work surface was 59 inches. One emitter was positive and the other negative.

Shielding/Discharge Test

This test for evaluating packaging materials was conducted with apparatus in accordance with published reports.[6,7] In brief, a capacitive sensor, comprising two 0.75-inch-diameter aluminum disks (0.06 inch thick) separated by 0.5 inch of acrylic plastic, was placed inside a 3 × 3 inch pouch of packaging material clamped between flat aluminum electrodes. See Fig. 2. In accordance with a published procedure,[8] a pulse of 1200 volts was applied from a 200 pF capacitor through a 400 kilohm resistor to the upper electrode, the lower electrode being grounded. The pulse picked up by the capacitive sensor was delivered to a storage oscilloscope. Results were conveniently expressed as "percent attenuation;" for example, if the pulse recorded by the sensor was 300 volts, the percent attenuation was (1200 − 300)(100)/1200 = 75. One edge of the pouch under test was always continous so that the material had electrical continuity from side to side.

Modified Shielding/Discharge Test

The capacitive sensor described above was provided with 4-inch leads ending in clips. MOSFETs, described above, were clipped with the substrate-case lead connected to the upper capacitor plate and the gate lead connected to the lower plate. The sensor, including the leads and the MOSFET, was placed inside an 8 × 10 inch pouch of packaging material resting on a grounded aluminum plate, and the charged probe of an Electro-Metrics

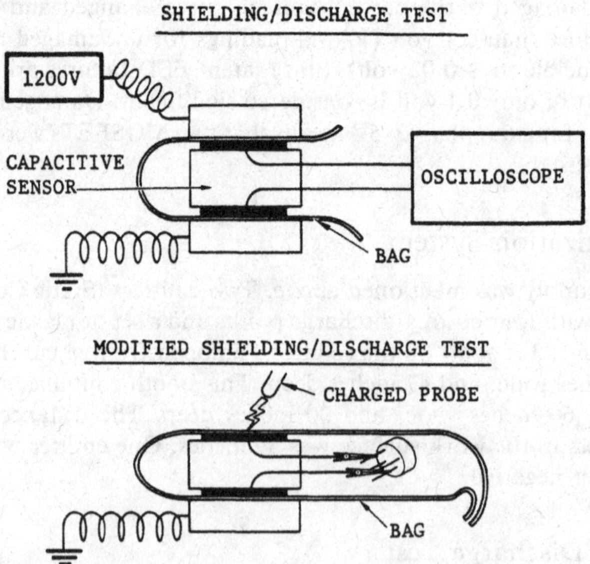

Fig. 2. Diagram of bag-testing procedures.

Model EDS-200 discharge unit (Model D-25, 150 ohms, 150 pF) was touched to the upper surface of the pouch (see Fig. 2). To ensure a worst-case condition, the packaging material was pulled tight against the sensor by taping down the edges of the pouch.

The probe of the discharge unit was intended to represent a statically-charged person's finger touching a bag containing ESDS items and lying on a conductive surface. In a few cases an actual finger was used.

Hazards of Charges at the Work Station

Bench Tops

In discussing bench tops we are concerned not with the top holding a charge but with its ability to drain charges from objects placed on it.

Static-dissipative tops (surface resisitivity 10^5–10^9 ohms/square) are favorably discussed in DoD-HDBK-263 and have been suggested as ideal.[9] Antistatic tops may drain charges too slowly, as will be seen below, while conductive tops are an electrical safety hazard and might damage charged ESDS items by discharging them too rapidly.[9,10] (See also DoD-HDBK-263 and the discussion below.)

A top may have a conductive sublayer without being satisfactory, because without sufficient surface conductivity the charge will be merely voltage-suppressed and not drained.[11] In our tests, conductive or antistatic tote boxes charged to ±10,000 volts drained incompletely (e.g., 8000 volts remained on an antistatic box) after resting 1 minute on an antistatic top (10^{11} ohms/square) but drained completely (< 50 volts remained) on a static-dissipative top (10^9 ohms/square). One successful construction for a static-dissipative top is, in our expereince, a high-pressure laminate with linen fabric in the surface layer and conductive carbon paper (grounded through 1 megohm) as the sublayer.[12] The solvent resistance of this laminate is reported to be very good,[12] and in our tests 50 daily rinses with methyl ethyl ketone had no effect on surface resistivity or appearance.

Static-dissipative tops are safer than conductive tops for operators, but can static-dissipative tops also be safer for devices as suggested above? To answer this question, tests were conducted with the MOSFET board (Fig. 1). An operator was charged to various levels by contacting the dc power supply, and a second after releasing the supply he touched the lead of the MOSFET board to either a static-dissipative bench top or a conductive work surface represented by an aluminum sheet. Both surfaces were grounded through 1 megohm.

Results are given in Table 1. Both surfaces allowed damage to occur, but

Table 1. MOSFET Damage Caused by Charged Operator Touching Lead of MOSFET Board (Fig. 1) to Work Surfaces Grounded through 1 Megohm.

	MOSFET DAMAGE	
VOLTAGE	STATIC-DISSIPATIVE SURFACE (8×10^8 OHMS/SQUARE)	ALUMINUM SHEET
−100	0/5	1/5
−200	0.5	3/5 (2 shorted)
−300	1.5 (1 shorted)	3/5 (1 shorted)
−400	0/5	2/5 (0 shorted)
−500	2/5 (1 shorted)	3/5 (2 shorted)
−1000	2/5 (1 shorted)	5/5 (4 shorted)
Totals:	5/30 (3 shorted)	17/30 (9 shorted)

the static dissipative surface was less destructive. That a more resistive surface can be less damaging will be seen again in Table 2.

In conclusion, an antistatic top will not itself hold a charge, but it drains charges too slowly from objects placed on it. The best choice seems a static-dissipative top with a buried conductive layer and a securely mounted grounding lug.

Table 2. MOSFET Damage Caused by Charged Operator Touching Lead of MOSFET Board (Fig. 1) to Objects on Static-Dissipative Bench Top (8×10^8 ohms/square) Grounded through 1 Megohm.

	MOSFET DAMAGE			
VOLTAGE	ANTISTATIC TOTE BOX	CONDUCTIVE TOTE BOX	ALUMINUM SHEET	PARTIALLY-TRANSPARENT BAG (11 × 14 INCH) WITH EXTERNAL METALLIZATION
−300	0/3	0/3	1/3	—
−500	0/3	0/3	2/3 (1 shorted)	—
−1,000	0/3	2/3 (1 shorted)	2/3 (1 shorted)	2/3 (0 shorted)
−2,000	0/3	3/3 (2 shorted)	—	3/3 (1 shorted)
−5,000	0/3	2/3 (1 shorted)	—	—
−10,000	0/3	3/3 (2 shorted)	—	—
Totals:	0/18	10/18 (6 shorted)		

Tote Boxes

Tote box selection is a controversial subject. Antistatic tote boxes can lose their "sweat layer" and should be periodically treated with an antistat solution.[1] Carbon-loaded plastic boxes, on the other hand, are permanently conductive but can slough conductive carbon particles. If sloughing is not considered a problem, one question remains: Are carbon-loaded boxes too conductive? The fear is that a charge on a conductive box might "zap" an ESDS item, whereas under the same conditions the charge would bleed off harmlessly from an antistatic box. Or, conversely, a charged ESDS item might discharge too rapidly to the "sparking surface" of a conductive box.[10,13]

We attempted to settle this question using the MOSFET board (Fig. 1). Either the operator or the tote box was charged, and the lead of the MOSFET board was touched to the box. Two boxes were tested: an antistatic box (surface resistivity 2×10^9 ohms/square at 48 percent relative humidity and 72°F) and a black conductive tote box (2.2×10^4 ohms/square at 32 volts). Both boxes were injection-molded by the same manufacturer and had the same dimensions (approx. 7 × 10 × 3 inches deep, with a 0.08-inch wall). Results are given in Tables 2 and 3, which also include an aluminum sheet (12 × 12 × 0.06 inch) as a reference. Table 2 also includes a static-shielding bag, discussed below. When the operator was charged, the box or sheet lay on a static-dissipative bench top grounded through 1 megohm; when the box or sheet was charged, it stood on insulat-

Table 3. MOSFET Damage Caused by Grounded Operator Touching Lead of MOSFET Board (Fig. 1) to Charged Objects.

	MOSFET DAMAGE		
VOLTAGE	ANTISTATIC TOTE BOX	CONDUCTIVE TOTE BOX	ALUMINUM SHEET
−300	0/3	0/3	2/3 (1 shorted)
−1,000	0/3	2/3 (1 shorted)	1/3 (1 shorted)
−2,000	2/3 (2 shorted)	3/3 (3 shorted)	3/3 (3 shorted)
−3,000	2/3 (1 shorted)	2/3 (1 shorted)	3/3 (3 shorted)
Totals:	4/12 (3 shorted)	7/12 (5 shorted)	9/12 (8 shorted)

ing stand-offs (plastic beakers) and the operator was grounded through 1 megohm with a wrist strap. The tests were conducted at 68 percent relative humidity and 72°F.

The result was that the antistatic tote box was clearly less damaging than the conductive box when the operator was charged but only slightly less damaging when the box was charged. An explanation is that when the box was charged (Table 3) induction became the dominant factor; that is, MOSFETs were damaged by capacitive coupling when the lead of the MOSFET board penetrated the field of the charged box. On the other hand, when the operator was charged (Table 2) the controlling factor was the resistance of the surface being touched by the lead.

The conclusion is that antistatic boxes are less of a discharge hazard than conductive boxes if handling techniques are poor. With good operator discipline, conductive boxes are safe and have the advantage of not relying on a fugitive antistatic additive; however, the user must assure himself that sloughing of conductive particles is not a problem.

In our assembly operations, modules are placed in Faraday-cage (foil-containing laminate) bags, and then the bags are placed in tote boxes. When a tote box is received at a work station, the operator removes the bag and sets the tote box aside before removing the module. Thus, the tote box is not a threat no mattter what its electrical properties. Even ordinary plastic tote boxes could be used in this manner, but good practice calls for selection of conductive or antistatic materials in an ESD control program. The blanket rule is that ordinary plastics must be excluded from work stations.

Nonconductive Plastics

When a polyethylene film (0.004 inch) at −13,000 volts (apparent charge, as are all charges reported in this paper) lay flat on grounded aluminum foil, the apparent charge was reduced to −300 volts because of voltage suppression, and MOSFET damage by a grounded operator touching the MOSFET board lead to the plastic surface was 0/3. Similarly, for a vinyl sheet (0.058 inch) at −8000 volts the apparent charge fell to −300 volts and MOSFET damage was 0/5. The nonconductive plastics were unable to deliver their high charges, and the suppressed fields were too low to cause damage.

The conclusion is that fields, rather than charges per se, are the hazard in this case. Thus the MOSFET damage reported in Table 4 when charged nonconductors were touched would presumably have been caused by a close approach without contact.

Table 4. MOSFET Damage Caused by Fields from Various Charged Objects.

OBJECT	APPARENT CHARGE, VOLTS	DISTANCE, INCHES	MOSFET DAMAGE
Butyrate Tote Box Lid	+7000	2	3/3 (3 shorted)
		6	1/3 (0 shorted)
Vinyl Tote Box on Bench	+500	0 (touched)	1/3 (0 shorted)
Vinyl Chair	+500	0 (touched)	1/5 (1 shorted)
Plastic Identification Badge	−2000	3	1/4 (0 shorted)
Butyrate Screwdriver Handle	+1700	0 (touched)	2/3 (0 shorted)
Polyester Lab Coat	+1300	0 (touched)	2/5 (0 shorted)
Polyethylene Film	−10,000	1	2/2 (1 shorted)
Aluminum Sheet	−3300	4	3/3 (1 shorted)
Painted Metal	+100	0 (touched)	0/3

Operators

An operator's skin should always be grounded. However, a grounded operator can cause ESD damage, as seen in Tables 3 and 4, and is only the lesser of two evils versus an ungrounded operator. Good handling techniques are needed to protect ESDS items from grounded operators when charges on the items are unavoidable. For example, a module might be triboelectrically charged by spraying with a conformal coating. Then, if a grounded operator were to touch a contact a damaging discharge could occur as explained in the literature.[10,14] A solution to this problem in terms of handling techniques is for the operator to be conscious of the hazard and avoid touching leads or contacts of ESDS items.

A possibility is that an operator's fingers might be reduced from conductive to static-dissipative or antistatic surfaces by his wearing appropriate gloves, possibly cotton. This idea is especially applicable to static-producing assembly or rework operations such as grit blasting.

Hazards of Fields at the Work Station

Field Strength versus Distance

Table 4 gives typical data obtained when a grounded operator waved the MOSFET board (Fig. 1) in the vicinity of various charged objects. Rapid motion, however, seemed unnecessary; when the lead of the MOSFET board was brought within 1 inch of a plastic surface at −5000 volts at the slow speed of 5 inches/minute, and then withdrawn at the same speed, MOSFET damage was 1/2. Presumably a potential difference was slowly built up by capacitive coupling until the oxide layer of the device was ruptured.

Incidentally, it should be noted that a charge of −3000 volts on the MOSFET board itself, with both the antennas and the operator grounded, caused 2 of 3 MOSFETs to be damaged (1 shorted) by installation and removal from the board. Damage of components during insertion into highly charged printed circuit boards has been reported.[15]

Figure 3 shows MOSFET damage levels caused by a grounded operator (wearing a wrist strap with a 1-megohm resistor) waving the MOSFET board in static fields. The circles are for various plastic surfaces, and the triangles are for an isolated $12 \times 12 \times \frac{1}{16}$ inch aluminum sheet charged to various negative voltages by the dc power supply. The black data points represent MOSFET failures (at least 1/3), and the white points represent no failures (0/3). Assuming a point source, field strength is directly proportional to voltage on the surface and inversely proportional to the square of the distance from the surface. Therefore, V plotted against d^2, where V is voltage and d is distance from the surface, will give a straight line whose slope represents field strength. The same result is given by plotting $\sqrt{V}$ versus d, which was done for convenience. Of course, the source of the field is an area rather than a point, and the data are not precise enough or numerous enough to establish an exact threshold field strength to cause damage. However, the line in Fig. 3 illustrates the principle. The equation for this line (except near the origin) is $\sqrt{V} = Cd$, or $d = \sqrt{V}/C$, where C is a constant (1.8 in this case) and d is the minimum safe distance in inches from a surface at apparent voltage V. For example, if the charge were 3000 volts, the calculated minimum safe distance would be $\sqrt{3000}/1.8 = 30$ inches. Since no damage occurs below 50 volts, the line does not pass through the origin, and the threshold for damage by nonconductors was 300–500 volts.

This estimate of minimum hazardous field strength is, of course, worst-case because the long antennas of the MOSFET board served to "gather" static fields. At the other extreme was a "MOSFET module" in which the

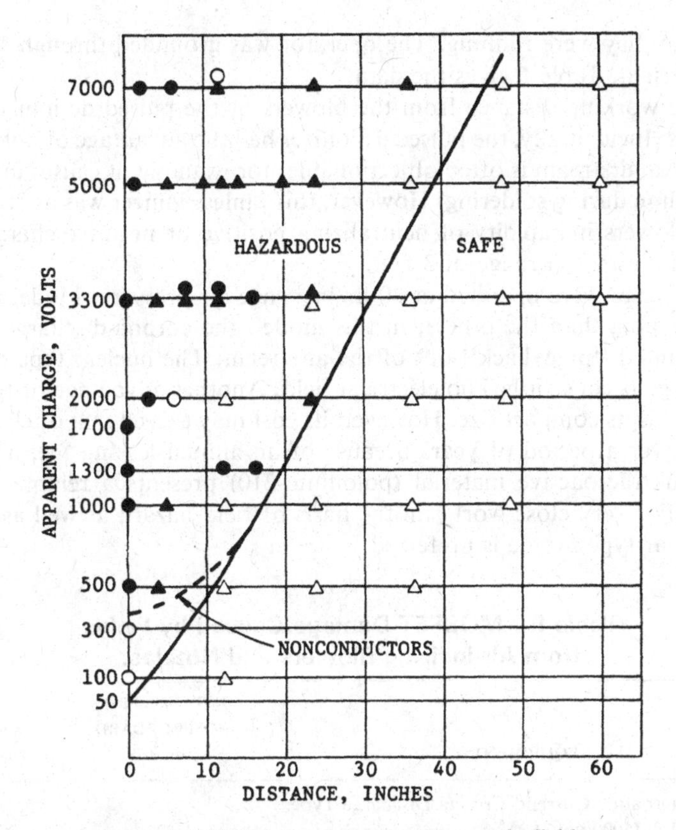

Fig. 3. MOSFET damage caused by static fields.

antennas (insulated wires) were sandwiched between aluminum ground planes while the projecting lead shown in Fig. 1 was eliminated. At 1 foot from a surface at −7000 volts, the "MOSFET module" gave MOSFET damage 0/5 versus 4/5 (two shorted) for the MOSFET board. However, the "MOSFET module" did allow damage (2/3, 0 shorted) at 2 inches from a surface at −7000 volts. Therefore, the constant *C* for the "MOSFET module" is more than 7 but less than 42. Real-life module assembly situations will lie between the two extremes represented by the MOSFET board and module.

Air Ionizers

A few ionized-air blowers and nozzles (guns) were evaluated for field hazards by waving the MOSFET board (Fig. 1) at various distances from their tips while they were running. The operator was grounded through 1 megohm as usual. Table 5 gives the data.

A safe working distance from the blowers or the pulsed dc ionizer was 10 inches. Incidentally, the pulsed dc ionizer had the advantage of not blowing air. An airstream is often objectionable; for example, it causes undesirable cooling during soldering. However, this fanless ionizer was as effective as the blowers in rapidity of neutralizing positive or negative charges on plastics in tests at a range of 3 feet.

The nozzles were operated at 30 lb/in.^{2} gage air pressure. Model C was less damaging than the others; in this model, the corona-discharge point was mounted "piggyback" out of the airstream. The nuclear type caused no damage because it has no electrical field. Another advantage of the nuclear type is its compact size. However, its cost may exceed that of electrical nozzles over a period of years because of an annual leasing fee, and the poisonous radioactive material (polonium-210) presents a remote safety hazard. For very close work, on the basis of field hazard as well as bulk, the nuclear-type nozzle is preferred.

Table 5. MOSFET Damage Caused by Fields from Air-Ionizing Blowers and Nozzles.

EQUIPMENT	DISTANCE FROM MOSFET BOARD, INCHES	MOSFET DAMAGE
Blower Alternating Current, Corona Discharge Type:		
Model A (5000 volts)	2	2/3 (1 shorted)
	4	0/5
	6	0/5
Model B (5000 volts)	4	4/10 (2 shorted)
	10	0/10
Ionizer, Pulsed Direct Current, Corona Discharge Type, 10,000 volts (Balanced)	10	0/10
Ionized Air Nozzles, Corona Discharge Type:		
Model A (4000 volts)	1	2/5 (0 shorted)
	2	0/5
Model B (4000 volts)	1	3/5 (1 shorted)
	2	1/3 (0 shorted)
	3	0/5
Model C (7000 volts)	1	0/5
Ionized Air Nozzle, Nuclear Type	1	0/5

The conclusion is that electrical air ionizers do not present a field hazard if the work is kept 10 inches or more away from them. However, when nozzles must be held very close to ESDS items, as in cleaning a module by blowing off particles, the nuclear type is required.

Room Ionization System

Reasons for selecting room ionization as opposed to local ionization—or for using ionization at all—are beyond the scope of this paper. We are concerned here only with certain aspects of a specific dc room ionization system.

The equipment which we tested is meant to be operated with the positive and negative emitters at similar voltages so that there is little net field strength or ion imbalance at the workbench level 5 feet or more beneath the emitters. However, if the system were accidentally unbalanced would an electrically isolated conductor on the bench become sufficiently charged to damage ESDS items touched to it?

To answer this question, tests were conducted in the booth described above under "Test Methods" using only one of the two emitters. The bench surface was static-dissipative as described under "Bench Tops," above. An aluminum sheet (12 × 12 × $\frac{1}{16}$ inch) was isolated from the bench surface by two different thicknesses of nonconductive plastic (bubble-wrap) as listed in Table 6. Thus, a capacitor was produced by the aluminum sheet and the carbon-paper sublayer of the bench; when the two were separated by $\frac{1}{4}$ inch, for example, the calculated capacitance of the aluminum sheet was 129 pF. The voltage on the aluminum sheet was measured with an Electrostatic Field Meter Model 970 from Static Control Services; in this case the field meter usually employed was not sensitive enough. Table 6 gives the data.

The system is normally operated at about 8200 volts; a slight increase, to perhaps 9000 volts, may be required at low humidity. Therefore, Table 6 indicates that even if all the emitters of one polarity should fail completely there will not be dangerous charging of isolated conductors lying on the bench. In conclusion, worst-case testing showed no hazard.

Having gained confidence that the room ionization system would not be part of the ESD problem, we tested its efficacy at neutralizing charges on plastics. Table 7 shows that neutralization of charges at the bench level was reasonably rapid for a variety of materials; the relative humidity during this test was 56 percent at 74°F. Even with the system somewhat unbalanced (15,000 positive ions/ml, 22,000 negative ions/ml), results were similar. To check the effect of humidity, the booth was sealed and the air inside it was partially dried with a desiccant. At 10–20 percent relative humidity, the decay rates of the materials in Table 7 remained rapid.

Table 6. MOSFET Damage Caused by an Unbalanced DC Room Ionization System.

EMITTER VOLTAGE*	DISTANCE FROM ALUMINUM SHEET TO BENCHTOP, IN.	OBSERVED VOLTAGE ON ALUMINUM SHEET	MOSFET DAMAGE
+17,000	4	+240	2/3 (1 shorted)
As required to charge aluminum sheet	4	+120	1/3 (0 shorted)
As required to charge sluminum sheet	4	+80	1/5 (0 shorted)
As required to charge aluminum sheet	1/4	+120	2/5 (1 shorted)
As required to charge aluminum sheet	1/4	+80	0/5
As required to charge aluminum sheet	1/4	+50	1/10 (0 shorted)
+8200	1/4	<10	0/10
−8200	1/4	<10	0/10

*Only one emitter was operating.

Another question was the effect of position of charged objects under the emitters; it was feared, for example, that a positively charged material under a positive emitter might receive relatively few negative ions and so exhibit a slow decay rate. Therefore a test was run with a postively charged plastic and a negatively charged plastic at 51 percent relative humidity and 73°F. Table 8 shows that charge decay rates remained far more rapid at any

Table 7. Charge Decay Rates with or without Room Ionization.

MATERIAL	INITIAL CHARGE, VOLTS	TIME, MINUTES, TO DECAY TO 500 VOLTS WITH SYSTEM ON*	RESIDUAL VOLTAGE AT 5 MINUTES WITH SYSTEM OFF**
Polystyrene Foam	+5000	1.5	+1200
Polyethylene Film	−500	1.8	−4800
Vinyl Sheet	−5000	2.5	−800
Carbon-Loaded Polyethylene (Conductive)	+5000	2.5	+3200
Aluminum Foil	+5000	1.8	+3300

*15,000 positive ions/ml, 13,000 negative ions/ml.
**<100 positive ions/ml, 500 negative ions/ml.

Table 8. Effect of Position Under Room-Ionization Emitters on Charge Decay Rates for Plastics.

MATERIAL	INITIAL CHARGE, VOLTS	POSITION ON BENCH SURFACE	PERCENT OF INITIAL CHARGE		
			5 MINUTES	10 MINUTES	20 MINUTES
Polyethylene Bubble-Wrap	−1700	Under Positive Emitter	3	—	—
		Under Negative Emitter	2	—	—
		Between Emitters	15	9	<1
		System Off	88	82	59
Butyrate Tote Box Lid	+15,000	Under Positive Emitter	17	11	3
		Under Negative Emitter	3	<1	—
		Between Emitters	4	<1	—
		System Off	93	87	80

position with the system on than when it was off. Tests at commercial installations of this and a competitive system gave better uniformity, presumably because a grid of emitters covered the ceiling instead of only two emitters being confined to a booth. Of course, perfect uniformity of decay rate across the bench is unecessary.

It was desired to compare the effectiveness of the room ionization system with that of a typical bench-model ionized-air blower (as in Table 5). Polyethylene bubble-wrap triboelectrically charged to −10,000 volts was found to lose half its charge after a 1-minute exposure (at 60 percent relative humidity and 74°F) either to the emitters at ±8200 volts at bench level or to the blower at a distance of 5 feet.

In conclusion, the room ionization system tested did not present a field or charge hazard at the bench level and was indeed effective in removing standing charges on nonconductors such as plastic packaging materials. Fig. 4 (for 49 percent relative humidity and 73°F), is a typical example of the utility of room ionization in removing a stubborn charge on plastic; similar curves appear in a published report.[16] Note that the electrical field of a charged item attracts ions to cause neutralization. Therefore, a thin sheet of plastic which lies flat on a workbench and has a largely collapsed field

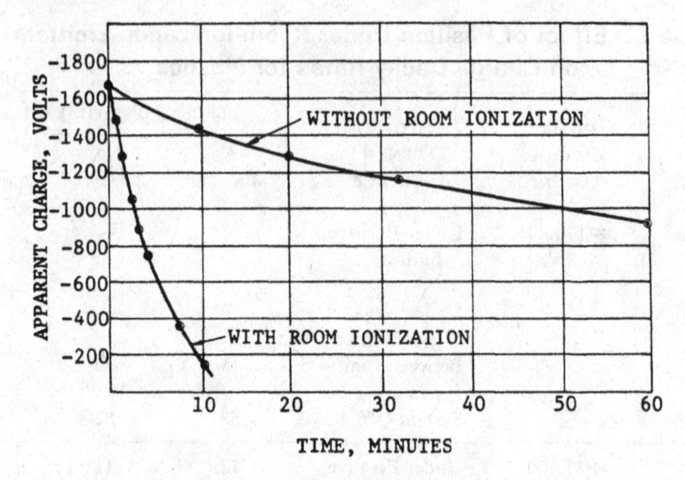

Fig. 4. Charge decay rates for polyethlene bubble-wrap.

will be neutralized only slowly and incompletely. The bubble-wrap in our test was $\frac{1}{2}$ inch thick and was folded over on itself so that the measured surface was elevated about 1 inch.

Assembly Operations

Many static-producing manufacturing operations are listed in DoD-HDBK-263 and a Navy training manual,[17] but only two will be discussed here: peeling masking tape from a roll and grit-blasting a module to remove the coating from components for rework.

A conductive masking tape would have virtually no static field because of charge drainage and voltage suppression. Our tests showed that a commercial aluminum foil/fiber tape had sufficient conformability and tear resistance for masking, but its acrylic adhesive released incompletely after oven-baking of the masked parts at 150°F. Therefore, the only deficiency of this tape might be overcome by sustituting a silicone adhesive, with good release, for the unsatisfactory acrylic.

A Micro Blaster (Comco, Inc) was found to create a charge of +700 volts on FR-4 circuit board material when sodium bicarbonate "grit" was sprayed against the surface using 10 lb/in.^{2} gage air pressure. By adding a "piggyback" corona-discharge ionizer (Simco P/N 4100034) operating at 7000 volts, with the point 0.6 inch from the spray nozzle, the charge developed was reduced to only +60 volts. This charge is well below the danger level for nonconductors (Fig. 2).

Operators

A properly grounded operator has no charge on his skin. However, arm hair has been reported to develop up to 900 volts, even at 50 percent relative humidity, when shirt sleeves are rolled up; at least one major corporation considers this a serious static problem.[18] Note that in our worst-case tests as little as 500 volts (on vinyl plastic, Table 4) was sufficient to damage a MOSFET. Obviously, head hair which dangles and might touch ESDS items is also a hazard.

A synthetic-fiber smock can carry damaging fields, as seen in Table 4, and DoD-HDBK-263 makes the point that clothing should never touch ESDS items. An "antistatic" smock is at least a partial solution, but it should not contain stainless-steel fibers which might fall onto circuit boards and cause shorts.[19] The operator's chair, if nonconductive, is also a hazard (Table 4).

The general approach to the problem of unavoidable static fields on an operator's hair or clothing—or on anything in the work station—is for him or her not to bring ESDS items near charged surfaces unnecessarily. In other words, do not look for trouble. In addition, head hair can be tied back, smock sleeves rolled up (with the proviso that ESDS items are kept away from hair on the forearms), etc.

ESD-Protective Equipment and Materials

Our discussion will be limited to two items: wrist straps and bags.

Wrist Straps

Various designs were tested. Bead chains gave intermittent contact but succeeded in protecting MOSFETs when the operator charged himself by shuffling his feet on a carpet and touched the lead of the MOSFET board to ground. The main objection to bead chains is their tendency to ride up over sleeves and lose contact with the skin. Any design which presses grounded metal snugly against the skin seems adequate; this includes expanding stainless-steel watchband types or designs in which a metal "wristwatch" element is held against the skin by an elastic band. However, wrist straps which depend on the conductivity of carbon-loaded plastic are suspect,[20] and a conductive fabric band shed steel fibers up to 0.09 inch long in our tests.

Faraday-Cage Bags

It is well recognized that ESD-protective bags must guard against (1) internal triboelectric charging and (2) external static fields/discharges.[9]

Except for polyethylene and antistatic polyethylene, the bags in Tables 9 and 10 all provide some shielding against external fields. In terms of the pulse attenuation test (Table 9), 93 percent attenuation must be sufficient because the foil-containing laminate bag is an excellent Faraday cage.[21]

The high values given by a partially transparent, externally metallized bag (99.7 percent) and a volume-conductive bag (97 percent) are artifacts, in a way, because conductive material "shorts out" the upper and lower plates in the test; wrapping these bags in antistatic polyethylene lowered their pulse attenuation to the 92–93 percent range. By this test it would seem that carbon-loaded polyolefin is as good a Faraday cage as aluminum foil, but such is not the case; discharges from a person's finger can damage ESDS items inside conductive plastic bags.[3,21]

A more significant test is the device-in-bag procedure of Table 10. The conditions were certainly worst-case because a very low resistance (150 ohms) was used; a resistance of 1500 ohms is specified in the human ESD model of DoD-HDBK-263, and a realistic estimate for human resistance is said to be 350 kilohms.[6] However, the foil-containing laminate bag survived even this test.

MOSFETs in a partially transparent, externally metallized bag were damaged by a discharge from the finger of a statically charged person (Table

Table 9. Shielding/Discharge Test.

BAG[1,21]	PERCENT PULSE ATTENUATION
Antistatic Polyethylene	75
Carbon-Loaded Polyolefin (Conductive)	97
With Conductive Ink Grid, A	78
With Conductive Ind Grid, B	50
Partially Transparent with External Metallization	99.8
Same, but Handled (Metallization Cracked on Crease)	59
Partially Transparent with "Buried" Metallization, A	39
Partially Transparent with "Buried" Metallization, B	87
Laminate Containing Aluminum Foil, with Antistatic Polyethylene Lining	93
Antistatic Polyethylene Bags Inside of and Outside of Carbon-Loaded Polyolefin Bag	92
Antistatic Polyethylene Bag Outside of Partially Transparent Bag with External Metallization	93

Table 10. Modified Shielding/Discharge Test.

BAG (AS IN TABLE 9)	VOLTAGE	MOSFET DAMAGE
Carbon-Loaded Polyethylene (Conductive)	+5000	3/3 (2 shorted)
With Conductive Ink Grid, A	+5000	1/3 (1 shorted)
Partially Transparent with External Metallization	+5000	3/3 (3 shorted)
Partially Transparent with "Buried" Metallization, A	+5000	1/3 (1 shorted)
Laminated Containing Aluminum Foil, with Antistatic	+5000	0/5
Polyethylene Lining	+10,000	0/5
Partially Transparent with External Metallization**	−8000 (est)*	2/3 (0 shorted)
Same, but Handled***	−8000 (est)*	2/3 (2 shorted)
Partially Transparent with "Buried" Metallization, A	−8000 (est)*	3/3 (0 shorted)

*The EDS-200 probe was replaced by the finger of a person triboelectrically charged by walking on a carpet at 24 % relative humidity and 73°F.
**Surface resistivity between sides = 256 ohms/square at 0.2 volt.
***Surface resistivity between sides = 2×10^6 ohms/square at 100 volts.

10), and the spark burned off a small area of metal. Published data[6] seem to disagree, but conditions differed somewhat from ours; for one thing, the capacitive probe had no leads as ours did. Also, the bag may not have been pulled tightly over the upper plate of the probe as it was on our test. As expected, MOSFET damage was worse (Table 10) when the externally metallized bag had a low-conductivity crease caused by cracking of the metallization by handling. Also note that externally metallized bags, being good conductors, can participate in discharges at the work station (Table 2).

The conclusion, which we had reached earlier,[21] is that only foil-containing laminate bags provide absolute protection against worst-case fields and discharges. A suitable commercial product is a three-layer laminate with aluminum foil (0.00035 inch thick) sandwiched between antistat-treated spun-bonded polyethylene on the outside and antistatic polyethylene on the inside. Note that MIL-B-81705, Type I, is a foil-containing laminate but has an unsatisfactory liner which does not prevent internal triboelectric charging.[4,9]

Equipment/Material Selection and Handling Techniques

One school of thought recommends exclusion of conductive materials[13] while another, the "conductive approach," visualizes most objects in the work station as conductive, grounded, and therefore at zero potential.[22]

Either approach, or a mixture of the two approaches, will be effective with proper maintenance of materials (e.g., periodic treatment of antistatic surfaces with topical antistat) and handling techniques which accommodate the limitations of the equipment/materials. Table 11 illustrates the principle of complementary materials and techniques.

The fact is that ESD-protective materials must be used correctly to be effective.[23] In other words, operators are more important than equipment.

Table 11. Complementary Equipment/Materials and Techniques.

ESD HAZARD	EQUIPMENT/MATERIAL	TECHNIQUES
Bench top	Static-dissipative laminate	Avoid unnecessarily touching any surface, even this, with leads of ESDS items.
Ordinary plastics	Ionization equipment, topical antistat	Identify and exclude if possible; treat with antistat; use local or room ionization.
Chair	If possible, use conductive chair on grounded conductive surface.	Don't closely approach with ESDS items.
Tote box	Conductive	Be aware of possible sloughing.
Tote box	Antistatic	Periodically treat with antistat solution.
Tote box	Either type	Ensure grounding when on work surface; avoid touching with ESDS items; place ESDS items in Faraday cage, such as foil-containing laminate bag, before adding to box; keep box to one side when working on ESDS items.
Operator's skin	Wrist strap	Always use strap and check its resistance often.
Operator's arm hair and head hair	None	Tie head hair back; don't closely approach hair with ESDS items.
Operator's clothing	Antistatic smocks	Don't closely approach with ESDS items.
Unavoidable static fields (eg, from electrical equipment)	Field meter	Avoid in accordance with Figure 3.
Static fields and discharges away from work station	Foil-containing laminate bags or other Faraday cages	Never remove ESDS items from Faraday cage except at static-safe work station.

For example, a grounded conductive chair would be more dangerous to ESDS items than an ordinary chair if a careless operator without a wrist strap caused discharges by brushing device leads against the upholstery. Conversely, a very skilled operator can protect ESDS items with only the bare essentials of equipment, including a wrist strap, because he knows where the dangers lie. Therefore, operator training is essential for ESD control programs as prescribed in DoD-STD-1686. A wide range of equipment/materials is described in DoD-HDBK-263, and it is up to the user to integrate these with operator disciplines in a unified approach.

In our program, handling techniques accommodate a mixture of conductive, static-dissipative, and antistatic materials. Some basic rules are: (1) Never touch ESDS device leads with the fingers unnecessarily, and keep ESDS items away from all sufaces except the bench top and the assembly to which the item is being attached. (2) Keep ESDS items in Faraday cages, such as foil-containing laminate bags, when away from the static-safe work station. (3) Clear the immediate work area of all but essential objects. A tote box might fail to drain completely and bear a charge—but if kept to one side it will be harmless. (4) Be sure to ground conductors, such as soldering iron tips, which touch leads or contacts and can discharge directly into ESDS devices. (5) Recognize the limitations of supplementary protective measures such as humidification or ionization and do not be overly dependent on them. Humidification merely lessens ESD hazards, and ionized air requries time to neutralize charges or cannot neutralize them at all insofar as the static field is collapsed by proximity to a ground plane. (6) Check wrist straps frequently. Various resistance testers are on the market. (7) Check with a field meter for static fields in the work station. For example, a nearby window might generate a high enough charge from dry wind blowing over it to create a dangerous field in the work area. If fields are unavoidable, apply the equation $d = \sqrt{V}/1.8$ as explained in the text above. (8) Be conscious that any assembly operation, such as peeling tape from a roll, creates static charges. Deal with these as required; for example, pull masking tape off the roll slowly and "wash" it in ionized air before applying it to contacts on a circuit board. (9) Remember that you, the operator, are a primary hazard even though grounded because you can participate in discharges. (10) Be scrupulous in exluding common plastics, which may be personal items such as sandwich bags, from the work station. As a last resort, ordinary plastics which definitely cannot be excluded must be treated with topical antistat.

In general, equipment and materials must be selected carefully from the gamut of products available, and then handling techniques must be adapted to the products chosen.

Conclusions

1. Tests confirmed that static-dissipative bench tops are preferred to conductive surfaces not only for personnel safety reasons but because a slower discharge rate can reduce damage to ESDS items.

2. A charged conductive tote box is only slightly more hazardous than a charged antistatic box when an ESDS item enters the static field of the box. But when boxes rest on a grounded static-dissipative work surface, a conductive box participates much more readily in harmful discharges than does an antistatic box. Both types of boxes have their pros and cons, and either is suitable if used properly.

3. Statically charged nonconductive plastics can present a field hazard when the apparent charge is as little as 500 volts. However, a charge of as much as 13,000 volts is not deliverable to an ESDS item by contact (discharge) when the field is suppressed to 300 volts apparent charge by proximity to a ground plane.

4. Under worst-case conditions, using a MOSFET with long "antennas," a rough equation for the minimum safe distance d (in inches) from a surface charged to voltage V is: $d = \sqrt{V}/1.8$. When the antennas are shielded, as they often are in practice on circuit-board modules, the constant in this equation rises from 1.8 to at least 7.

5. Based on the above data, basic rules are: (1) Ground all conductors so that none can have a charge of over 50 volts. (2) Check with a field meter and allow no more than 300 volts apparent charge on nonconductors such as ordinary plastics. (3) If static fields are unavoidable, keep ESDS items d inches away from charged surfaces according to the worst case equation $d = \sqrt{V}/1.8$.

6. Electrical air-ionizing blowers or nozzles were found not to present a field hazard if ESDS items are kept at least 10 inches away from the corona-discharge points. For close work, nuclear-type air nozzles are indicated.

7. A pulsed dc benchtop ionizer, with no fan, was found to be as effective as conventional ac blowers. The lack of an airstream is advantageous. (Note: Slow-pulsed ionizers later proved hazardous; see "Ionization" in Chapter 2.)

8. A room ionization system with dc emitters proved to be effective—about equal to a typical bench-model blower at 5 feet—and relatively uniform regardless of the position of charged objects on the bench surface under the emitters. At worst-case conditions (emitter of only one polarity operating), an isolated conductor at bench level (capacitance 129 pF) was not charged sufficiently to damage MOSFETs.

9. The possibility of a conductive masking tape of foil/fiber construction

is suggested. The aluminum foil prevents a static field by charge drainage and voltage suppression.

10. A corona-discharge point mounted above a small-scale grit blaster reduced the charge developed on circuit-board material from 700 volts to the harmless level of 60 volts.

11. A bead-chain wrist strap is effective but tends to ride up over sleeves. Snug-fitting straps in which metal contacts the skin are recommended.

12. In our tests, MOSFETs were damaged in partially transparent metallized bags, under worst-case conditions, when the bags were touched by an operator charged to an estimated 8000 volts by walking on a carpet. In contrast, MOSFETs in a foil-containing laminate bag were unaffected by a 10,000-volt discharge from a simulator with human-level capacitance but three orders of magnitude less resistance than a person. Only foil-containing laminate bags are recommended for Farday-cage protection against worst-case fields and discharges.

13. A synthesis of complementary equipment/materials and handling techniques is necessary for an effective ESD control program in accordance with Dod-HDBK-1686. Examples are given.

14. Operators are more important than materials. Some basic operator disciplines are, in brief: (1) keep ESDS items away from all surfaces except the bench top and the assembly to which the item is being attached, (2) use Faraday-cage bags/containers, (3) clear the immediate work area of all but essential objects, (4) beware of charged conductors, (5) do not be overly dependent on supplementary protection such as humidification or ionization, (6) check wrist straps frequently, (7) check for static fields with a meter, (8) beware of static charges caused by assembly operations, (9) remember that a grounded operator can participate in harmful discharges, and (10) rigorously exclude common plastics.

(Note: Later research modified some of the above conclusions, as seen in the text of this book. For example, we became more cautious about ionization.)

References

1. J. M. Kolyer and W. E. Anderson, "Permanence of the Antistatic Property of Commercial Antistatic Bags and Tote Boxes," *Reliability Analysis Center EOS/ESD Symposium Proceedings,* pp. 87–94, Las Vegas, NV, 1983.
2. B. A. Unger, "Electrostatic Discharge Failures of Semiconductor Devices," Reliability Physics Symposium, April 1981.
3. J. R. Huntsman, D. M. Yenni, Jr., and G. E. Mueller, "Fundamental Requirements for Static Protective Containers," presented at 1980 NEPCON/West Conference, Anaheim, CA

4. D. M. Yenni, Jr., and J. R. Huntsman, "The Deficiencies in Military Specification MIL-B-81705: Considerations and a Simple Model for Static Protection," presented at the Reliability Analysis Center EOS/ESD Symposium, Denver, CO, 1979.
5. "Latent ESD Failures: A Reality," *Evaluation Engineering Magazine,* p. 80, April 1982.
6. J. R. Huntsman and D. M. Yenni, Jr., "Test Methods for Static Control Products," *Reliability Analysis Center EOS/ESD Symposium Proceedings,* pp. 94–109, Orlando, FL, 1982.
7. Electronics Industry Association Interim Standard No. 5, January 1983.
8. "Electromagnetic Shielding Effectiveness," technical information sheet from the Bemis Co., Inc., ESD Protective Materials Dept., undated, received 1983.
9. N. B. Fuqua and R. C. Walker, "ESD Controls Study, Final Report," Reliability Analysis Center No. 01115-30-2, September 1981.
10. B. Unger, R. Chemelli, P. Bossard, and M. Hudock, "Evaluation of Integrated Circuit Shipping Tubes," *Reliability Analysis Center EOS/ESD Symposium Proceedings,* pp. 57–64, Las Vegas, NV, 1981.
11. J. R. Huntsman and D. M. Yenni, Jr., "Charge Drainage vs. Voltage Suppression by Static Control Table Tops," *Evaluation Engineering Magazine,* March 1982.
12. J. R. Mileham and N. I. Safeer, "Selection of Static Eliminating Decorative Table Top Mats or Laminates," presented at 1984 NEPCON/West Conference, Anaheim, CA. The material we tested is No. 6 in the tables of this reference.
13. D. C. Anderson, "ESD Control: To Prevent the Spark that Kills," *Evaluation Engineering Magazine,* pp. 120–131, July 1984.
14. R. G. Chemelli, B. A. Unger, and P. R. Bossard, "ESD by Static Induction," *Reliability Analysis Center EOS/ESD Symposium Proceedings,* pp. 29–35, Las Vegas, NV, 1983.
15. J. E. Berry, "Static Control in Bare Board Testing," *Electronic Packaging and Production,* pp. 161–162, August 1981.
16. C. F. Mykkanen and D. R. Blinde, "The Room Ionization System: An Alternative to 40 Percent RH," *Evaluation Engineering Magazine,* pp. 76–88, September 1983.
17. "Electrostatic Discharge Training Manual," NAVSEA SE 003-AA-TRN-010, Published by Direction of Commander, Naval Sea Systems Command.
18. "Arm Hair Pin-Pointed as New ESD Hazard," *Evaluation Engineering Magazine,* p. 70, May 1984.
19. GIDEP (Government-Industry Exchange Program) Alert No. D5-A-84-01, May 21, 1984.
20. GIDEP (Government-Industry Exchange Program) Alert No. MX-A-82-02, March 21, 1983.
21. J. M. Kolyer and W. E. Anderson, "Selection of Packaging Materials for Electrostatic Discharge-Sensitive Items," *Reliability Analysis Center EOS/ESD Symposium Proceedings,* pp. 75–84, Las Vegas, NV, 1981.
22. R. Euker, "ESD in Integrated Circuit Assembly," *Static Digest,* published by Static Control Systems Division, 3M Co., July 1983.
23. G. E. Hansel, "The Production Operator: Weak Link or Warrior in the ESD Battle?" *Reliability Analysis Center EOS/ESD Symposium Proceedings,* pp. 12–16, Las Vegas, NV, 1983.

Paper No. 4

Presented at the 7th Annual Electrical Overstress/Electrostatic Discharge Symposium, Minneapolis, Minnesota, September 10-12, 1985, Sponsored by EOS/ESD Association and ITT Research Institute. *EOS/ESD Symposium Proceedings,* EOS-7, 1985, page 111.

PERFORATED FOIL BAGS: PARTIAL TRANSPARENCY AND EXCELLENT ESD PROTECTION

John M. Kolyer and William E. Anderson

Rockwell International Corporation
Autonetics Strategic Systems Division
Electronics Operations
3370 Miraloma Avenue
Anaheim, CA 92803-4192

Note: The following is a condensed version of this paper. See the *Symposium Proceedings* for the complete text.

Abstract

Commercial foil-containing laminate bags provide "Faraday-cage" protection for electrostatic discharge (ESD)-sensitive items but are opaque. However, foil may be perforated to give see-through capability if the metal is thick enough and the stamped holes have outwardly directed burrs which may act as "lightning rods" to intercept discharges. A wire screen serves the same purpose, with its high points presumably receiving the discharges. This principle is applicable to various other containers, e.g., tote boxes. Thus, partial transparency need not mean a sacrifice in ESD protection.

Introduction

Commercial "Faraday-cage" bags are intended to protect against electrostatic fields and discharges. A secondary goal is protection against EMI (electromagnetic interference), specifically at 1-10 GHz per MIL-B-81705, Type I. These bags fall into four classes: (1) foil-containing laminates which give good ESD and EMI protection but are completely opaque, (2) metallized bags in which the metallized layer is thin enough to allow partial transparency with a considerable sacrifice of ESD and EMI protection (also, external metallization can crack on handling, thus lowering shielding effectiveness), (3) bags with a grid of conductive ink that are fairly transparent but give poor ESD protection, and (4) metal-fiber-containing bags that are also fairly transparent but give poor ESD protection.

Therefore, our objective was a new bag laminate which combines the "best of both worlds" to a considerable extent: see-through capability along with good ESD protection and EMI protection in the 1-10 GHz (radar) range.

Materials Tested

Nine commercial ESD-protective bag materials were tested (Table 1): antistatic poly, carbon-loaded, ink grid, buried metallization A, external nickel, copper fibers, buried metallization B, buried metallization C, and foil laminate. Several of these products have been described and characterized by us.[1,2,3]

The other six materials in Table 1 were fabricated in the laboratory from antistatic polyethylene (6 mils) and either aluminum foil, perforated aluminum foil, or metallic screens.

Experimental Methods

The test conditions are listed in Table 2 and diagrammed in Figs. 1, 2, and 3.

Results

The data are summarized in Table 3. Note the excellent results given by screens. After publication of this paper a screen bag was commercialized; the aluminum screen was sandwiched between layers of antistatic polyethylene. Perforated foil was not as effective as screen and just as expensive, so it was never commercialized.

Conclusions

1. Metallized, carbon-loaded, or ink grid bags have been reported not to withstand a direct static charge above 25 kV. Our own tests indicate a much lower threshold for these materials. Note that personnel walking on a carpet at 10–20 percent relative humidity generate typically 35 kV.

2. Two independent groups of investigators have recommended (EOS/ESD Symposium, 1984) that highly ESD-sensitive items should always be protected by metal-foil bags.

Table 1. Bag Materials Tested.

DESIGNATION	THICKNESS MILS	TRANSPARENCY, %	DESCRIPTION
Antistatic Poly	6	87	Antistatic Polyethylene (MIL-B-81705, Type II)
Carbon-Loaded	4	0	Conductive, Carbon-Loaded Polyethylene
Ink Grid	4	68	Laminate with Buried Grid of Conductive Ink and Antistatic Liner
Buried Metallization A	3.5	74	Polyester + Vapor-Deposited Metal + Antistatic Polyethylene Liner
External Nickel	2.5	35	Vapor Deposited Nickel + Polyester + Antistatic Polyethylene Liner
Copper Fibers	3.7	78	Fine Copper Wires Dispersed in Polyethylene
Buried Metallization B	3.3	60	Resinous Protective Coating + Vapor-Deposited Aluminum + Carrier Film + Antistatic Polyethylene Liner
2-mil Perforated Foil	12	38	Perforated Foil Encapsulated in Antistatic Polyethylene
Buried Metallization C	3	60	Polyester + Vapor-Deposited Aluminum + Antistatic Polyethylene Liner
5-mil Perforated Foil	15	38	Perforated Foil (See Text) Encapsulated in Antistatic Polyethylene
Foil Laminate	10	0	Antistatic Spun-Bonded Polyethylene + 0.35-mil Aluminum Foil + Antistatic Polyethylene Liner
1.4-mil Foil	7.4	0	1.4-mil Aluminum Foil + 6-mil Antistatic Polyethylene Liner
5.6-mil Foil	11.6	0	Four layers of 1.4-mil Aluminum Foil + 6-mil Antistatic Polyethylene Liner
Aluminum Screen	23	59	Aluminum Screen (See Text) Sandwiched Between Sheets of 6-mil Antistatic Polyethylene
Copper Screen	35	41	Copper Screen (See Text) Sandwiched Between Sheets of 6-mil Antistatic Polyethylene

Table 2. Shielding/Discharge Test Conditions.

	TEST NUMBER							
	1	2	3	4	5	6	7	8
Capacitive Sensor	No	No	Yes	Yes	Yes	Yes	Yes	Yes
Volts	+25,000	About 35,000	+5000	−6000 to −8000	+10,000	About 35,000	+24,000	About 35,000
Resistance, ohms	0	Tesla Coil	150	Human	150	Tesla Coil	150	Tesla Coil
Capacitance, picofarads	167	Testa Coil	150	Human	185	Tesla Coil	185	Tesla Coil
Time, seconds	(Discharge)	30	(Discharge)	(Discharge)	(Discharge)	30	(Discharge)	30
Surface Under Bag	Nonconductive	Nonconductive	Grounded	Grounded	Grounded	Nonconductive	Grounded	Grounded

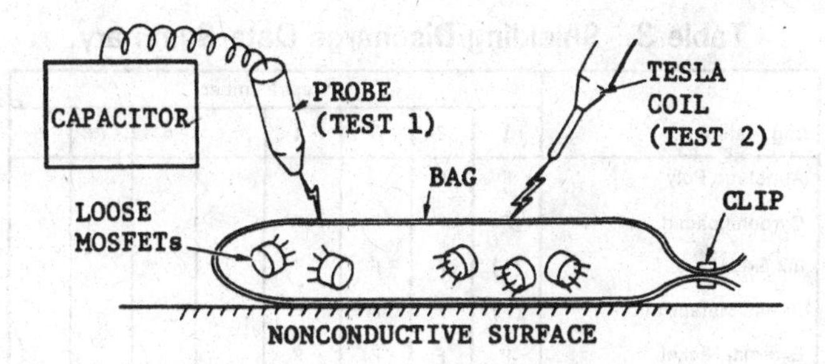

Fig. 1. Tests number 1 and 2 (Table 2).

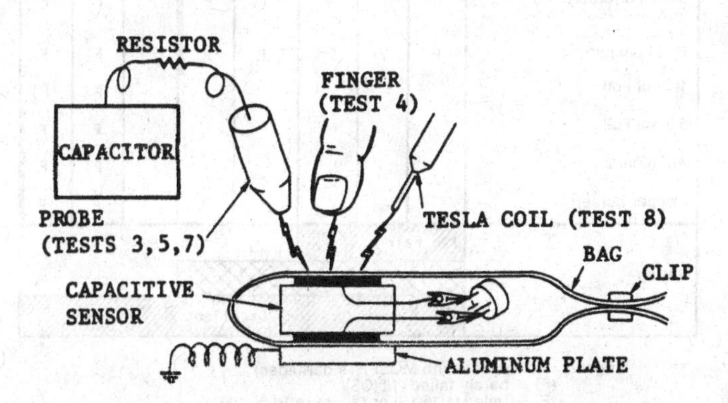

Fig. 2. Tests number 3, 4, 5, 7, and 8 (Table 2).

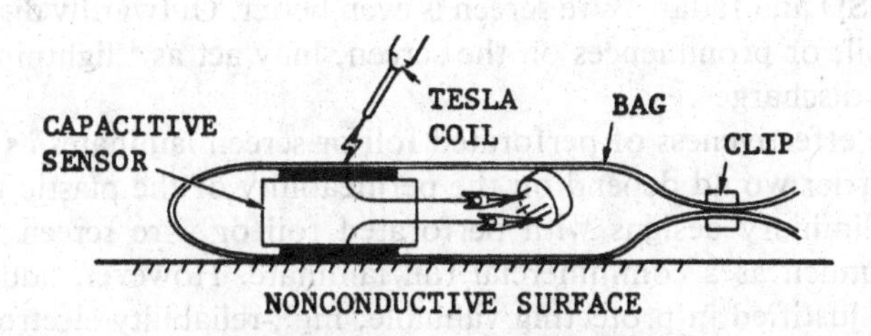

Fig. 3. Test number 6 (Table 2).

Table 3. Shielding/Discharge Data Summary.

Bag Material	Test Number 1	2	3	4	5	6	7	8
Antistatic Poly	F	F						
Carbon-Loaded	F	F	F	F				
Ink Grid	(F)	F	F	F				
Buried Metallization A	(F)	F	F	F				
External Nickel	P	F	F	F				
Copper Fibers	P	F	F	F				
Buried Metallization B	P	F	F	F				
2-mil Perforated Foil		P	(F)		F		F	
Buried Metallization C		P	P	(F)	F	F	F	F
5-mil Perforated Foil	P	P	P	P	P	F	F	F
Foil Laminate	P	P		P	P	(F)	F	F
1.4-mil Foil							P	(F)
5.6-mil Foil							P	P
Aluminum Screen							P	P
Copper Screen							P	P
Rating	Fair (tests 1–3)							
	Good (tests 1–5)							
	Excellent (tests 1–8)							

Legend: P = passed (no MOSFETs damaged)
(F) = barely failed (1/5(OS))
F = failed (1/15(1S) or failure ratio > 1/5)

3. We have now found that metal foil can be perforated and still shield against ESD and radar. Wire screen is even better. Outwardly directed burrs on the foil, or prominences on the screen, may act as "lightning rods" to intercept discharges.

4. The effectiveness of perforated foil or screen laminates as a moisture vapor barrier would depend on the permeability of the plastic film used.

5. Preliminary designs with perforated foil or wire screen would cost twice as much as a commmercial foil laminate. However, added expense might be justified in protecting valuable, high-reliability electronics.

References

1. J. M. Kolyer and W. E. Anderson, "Selection of Packaging Materials for Electrostatic Discharge-Sensitive Items," *Reliability Analysis Center EOS/ESD Symposium Proceedings,* pp. 75–84, Las Vegas, NV, 1981.
2. J. M. Kolyer and W. E. Anderson, "Permanence of the Antistatic Property of Commercial Antistatic Bags and Tote Boxes," *Reliability Analysis Center EOS/ESD Symposium Proceedings,* pp. 87–94, Las Vegas, NV, 1983.
3. J. M. Kolyer, W. E. Anderson, and D. E. Watson, "Hazards of Static Charges and Fields at the Work Station," *Reliability Analysis Center EOS/ESD Symposium Proceedings,* pp. 7–19, Philadelphia, PA, 1984.

Paper No. 5

Presented at the Electrical Overstress Exhibition, Anaheim, California, January 21, 1986. *Proceedings of the Technical Program, 1986 Electrical Overstress Exhibition,* page 34.

COST-EFFECTIVE METHODS OF TESTING/MONITORING WRIST STRAPS

John M. Kolyer and Donald E. Watson

Rockwell International Corporation
Autonetics Strategic Systems Division
Electronics Operations
3370 Miraloma Avenue
Anaheim, CA 92803

Abstract

An ungrounded operator is extremely dangerous to ESD (electrostatic discharge)-sensitive devices as illustrated by a test at high relative humidity (54 percent) in which the operator damaged transistors by rolling his chair. An ESD-protective smock did not help to protect devices. A maximum resistance from operator to ground of 10 megohms is recommended, and this value was exceeded in 3 of 36 tests under manufacturing conditions. Because of flexing of the cords and other abuse, wrist straps have a limited lifetime, but they are too expensive to scrap in a replacement cycle and instead must be monitored. Periodic monitoring is unsatisfactory because ESD-sensitive items will be exposed to an ungrounded operator for some period of time, even if only an hour, when a strap fails. Then material review action might be required on all hardware handled since the last successful test of that strap. Therefore, continuous monitors may be "the wave of the future." Commercial continous monitors work on the capacitance principle, but another method is a continuity check across the operator's skin, up and down a two-conductor cord, and in and out two separate ground connections. Prototype monitors of this two-conductor design are being tested in a production environment. High-cost items, e.g. $1000 modules, especially deserve continous monitoring.

Introduction

This paper is presented in the form of slides shown at the Electrical Overstress Exposition (EOE) in Anaheim, CA, January 21, 1986. Accompanying each slide is additional discussion and clarification as required.

Wrist straps have a limited lifetime[1] because of frequent flexing and other abuse. Since a time-based replacement cycle would scrap many good straps and be wasteful, monitoring is necessary. This monitoring is preferably continuous, because the failure of a wrist strap in a periodic check raises the questions: "How long ago did this strap fail?" and "Have products been damaged in the meantime?"

As shown in the following outline, the danger of a failed strap was demonstrated, simple methods for testing resistance were devised, straps in an assembly plant were checked, and means of monitoring were investigated.

OUTLINE

- DAMAGE CAUSED BY LACK OF WRIST STRAP
- RESISTANCE VS CHARGE ON OPERATOR
- ALLOWABLE RESISTANCE, OPERATOR TO GROUND
- RESISTANCE TEST METHODS
    - THROUGH STRAP
    - OPERATOR TO GROUND
    - EQUIPMENT
        - MEGOHMMETER
        - ZAPFLASH
- RESISTANCE TESTING DATA
- FACTORS IN RESISTANCE, OPERATOR TO GROUND
- MONITORING
    - PERIODIC VS CONTINUOUS
    - CONTINUOUS DESIGNS
        - CAPACITANCE TYPE
        - TWO-CONDUCTOR TYPE
        - ADVANTAGES OF TWO-CONDUCTOR TYPE
- CONCLUSIONS

The charge on the operator was read by means of a Trek Model 512 field meter aimed at an electrically isolated 1-ft^2 aluminum plate to which the operator was attached by a wrist strap. The temperature was 72°F.

The transistor was a Motorola 2N4351 MOSFET. The testing procedure and damage criteria have been described.[2]

DAMAGE CAUSED BY LACK OF WRIST STRAP

- GROUNDED OPERATOR IS LESSER OF TWO EVILS VS UNGROUNDED OPERATOR

  (KOLYER ET AL, EOS/ESD SYMPOSIUM 1984)

- UNGROUNDED OPERATOR IN ESD-PROTECTIVE SMOCK, 54% RELATIVE HUMIDITY

  - ROLLED CHAIR: 500 VOLTS, TRANSISTORS DAMAGED

  - SHUFFLED FEET: 1000 VOLTS, TRANSISTORS DAMAGED

A charged operator is a very great danger because he can damage ESD-sensitive devices by direct discharge or "injection" rather than indirectly by induction from a static field.

The published data[3] were obtained with an operator holding a probe attached to an electrometer and an oscilloscope. The Trek meter and 1-ft^2 plate were used as described under the preceding slide. Since 10 volts was our limit of detection, the result at 1 megohm agrees with that obtained by the more sophisticated method: a 1-megohm resistor in the wrist strap allows negligible voltage on the operator.

RESISTANCE VS CHARGE ON OPERATOR

RESISTANCE, OPERATOR TO GROUND, MEGOHMS	PEAK VOLTAGE ON OPERATOR	REFERENCE
1	<2	1
10	11	1
50	30	1
100	80	1
INFINITE (NO STRAP)	1550	1
1	<10	2
INFINITE (NO STRAP)	1100	2

1. HUNTSMAN AND YENNI, EOS/ESD SYMPOSIUM 1982
2. OUR DATA

ALLOWABLE RESISTANCE, OPERATOR TO GROUND

- FOR DEVICES SENSITIVE TO 50 VOLTS MIN: 50 MEGOHMS
- FOR DEVICES SENSITIVE TO 20 VOLTS MIN: 10 MEGOHMS
- CONSERVATIVE LIMIT FOR ALL DEVICES: 10 MEGOHMS

These thresholds are based on the data of the preceding slide. A charge of 11 volts on the operator is harmless to most devices at the present time; in the future this allowable resistance might have to be lowered. A 1-megohm resistor is presently standard for wrist straps.

As an added precaution, a static-limiting floor finish (UL288, Hanson-Loran Chemical Company) was used. The chair-rolling or foot-shuffling operator in the slide before last would have generated little charge with UL288. In the test, a conventional acrylic finish was on the tile floor.

A Beckman L-10 megohmeter at 10 volts was used. For quick tests, a Zapflash (see two slides below) is suitable.

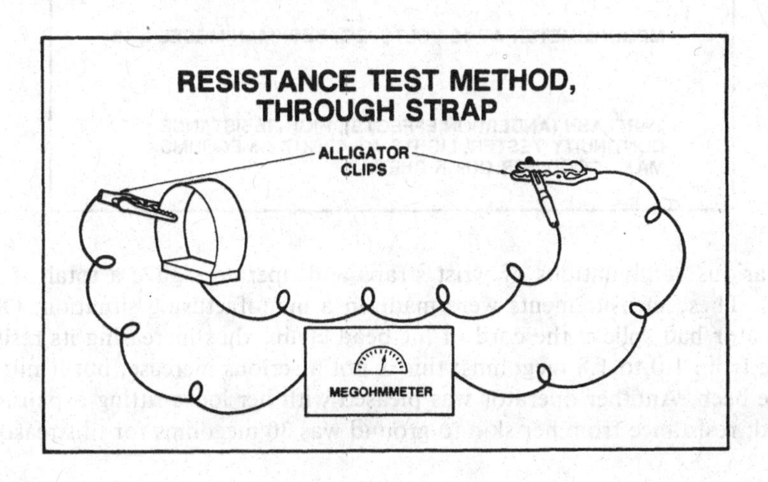

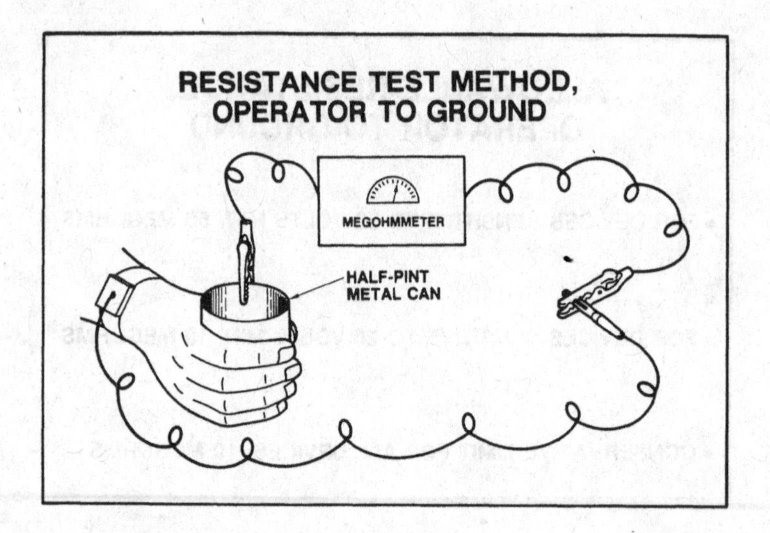

The same megohmeter at 10 volts was used. The metal can (tin-coated steel) is a convenient, hand-filling probe.

The Zapflash was tested with various resistance levels. At 1.0 megohm the light (indicating continuity) was very bright. At 3.0 megohms the light was relatively bright, but at 5.2 megohms it was dim and at 6.2 megohms it was barely visible.

RESISTANCE TEST METHODS, EQUIPMENT

- MEGOHMMETER AT 10 VOLTS, EG, BECKMAN MODEL L-10
- ZAPFLASH (ANDERSON EFFECTS), HIGH-RESISTANCE CONTINUITY TESTER, LIGHTS AT ABOUT 6 MEGOHMS MAX. GOOD FOR QUICK CHECKS

Various combinations of wrist straps and operators gave a total of 36 tests. These measurements were made in a manufacturing situation. One operator had spliced the cord of his bead chain, thus increasing its resistance from 1.0 to 1.8 megohms; this is not a serious increase, but it might have been. Another operator was pleased with her loose-fitting expansion band; resistance from her skin to ground was 30 megohms for this reason.

RESISTANCE TEST DATA

TYPE STRAP	NUMBER SAMPLES	NUMBER OPERATORS	RESISTANCE, MEGOHMS THROUGH STRAP	RESISTANCE, MEGOHMS OPERATOR TO GROUND
EXPANSION BAND (1)	5	7	1.00-1.05	1.0-9.0
EXPANSION BAND (2)	5	6	1.00-1.05	1.05-30
EXPANSION BAND (3)	1	2	1.0	2.0, 4.3
CLOTH BAND (CONDUCTIVE)	3	4	2.7-110	1.5-55
BEAD CHAIN	4	6	1.0-1.8*	1.5-40

*SPLICED CORD GAVE 1.8

These examples illustrate the need for operator discipline, especially in the absence of continous monitoring of the straps.

The three out-of-spec cases occurred in a program of periodic (weekly) checks and illustrate the desirability of continuous monitoring. In a periodic check, the operator may slide the band up his arm to tighten it or make some other adjustment to pass the test. An analogy is annual inspection of automobile exhaust emissions; once the driver has passed inspection, he might readjust his engine to exceed the allowable level of smog-producing gases.

FACTORS IN RESISTANCE, OPERATOR TO GROUND

- OVERSIZED EXPANSION BAND
- INCREASED RESISTANCE OF CLOTH BAND (SOILING)
- STRETCHING OF CLOTH BAND, CAUSING LOOSENESS
- SOILING OF BEAD CHAIN
- DRY SKIN
- IN 36 TESTS, RESISTANCE EXCEEDED 10 MEGOHMS IN 3 CASES: EXPANSION BAND (LOOSE FIT), CONDUCTIVE CLOTH BAND, BEAD CHAIN

PERIODIC VS CONTINUOUS MONITORING

- PERIODIC
  - ADVANTAGE: LOW EQUIPMENT COST
  - DISADVANTAGE: ESD-SENSITIVE DEVICES ENDANGERED FOR SOME TIME INTERVAL, EVEN IF ONLY AN HOUR
- CONTINUOUS
  - ADVANTAGES: DEVICES ALWAYS PROTECTED, WRIST STRAP QUALITY SECONDARY, LESS OPERATOR DISCIPLINE NEEDED
  - DISADVANTAGE: RELATIVE HIGH EQUIPMENT COST (BUT COST OF LABOR FOR PERIODIC MONITORING MUST BE SUBTRACTED)

CONTINUOUS MONITOR, CAPACITANCE TYPE

- CHECKS CAPACITANCE, EG, 150 pF
- ALARM FLASHES AT ALLOWABLE MAX OPERATOR-TO-GROUND RESISTANCE, EG, 3.5 MEGOHMS TOTAL
- USES CONVENTIONAL WRIST STRAP
- COMMERCIALLY AVAILABLE FROM MORE THAN ONE SUPPLIER

Commercial products include the Simco Wrist Strap Monitor Model M50A, the Ground Gard from Static Prevention, Incorporated, and the GM-1CT from Westek. Charleswater Products has a developmental product to be introduced early in 1986.

Note that in the absence of continuous monitoring, all hardware handled since the last successful test of a failed strap might be subject to a material review action.

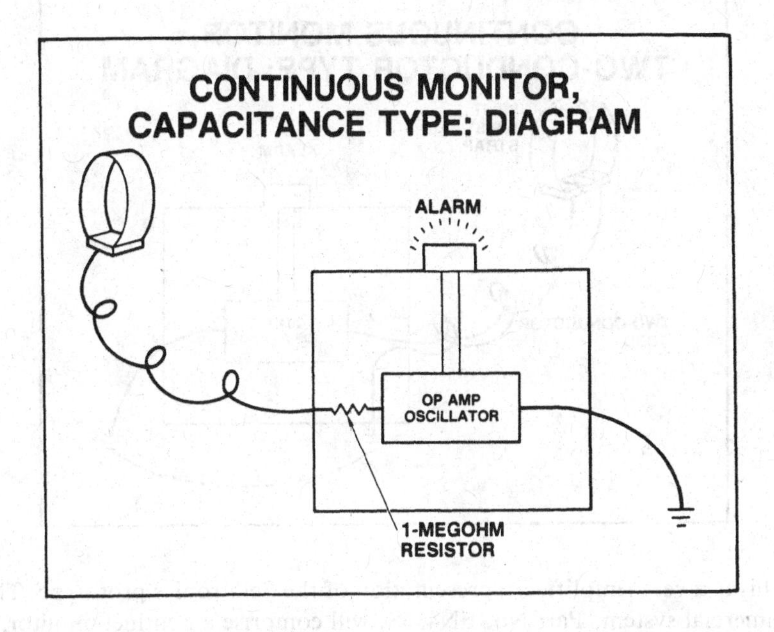

This is a very simplified representation of the Simco Wrist Strap Monitor Model M50A.

CONTINUOUS MONITOR, TWO-CONDUCTOR TYPE

- CHECKS LOOP ACROSS SKIN, INTO FIRST GROUND CONNECTION, AND OUT SECOND GROUND CONNECTION
- ALARM FLASHES AT 10 MEGOHMS, OPERATOR-TO-GROUND
- SPECIAL SPLIT-CONDUCTOR WRIST STRAP NEEDED
- PROTOTYPES TO BE TESTED IN PLANT

The product under evaluation is from Semtronics Corporation. Westek is developing their own model.

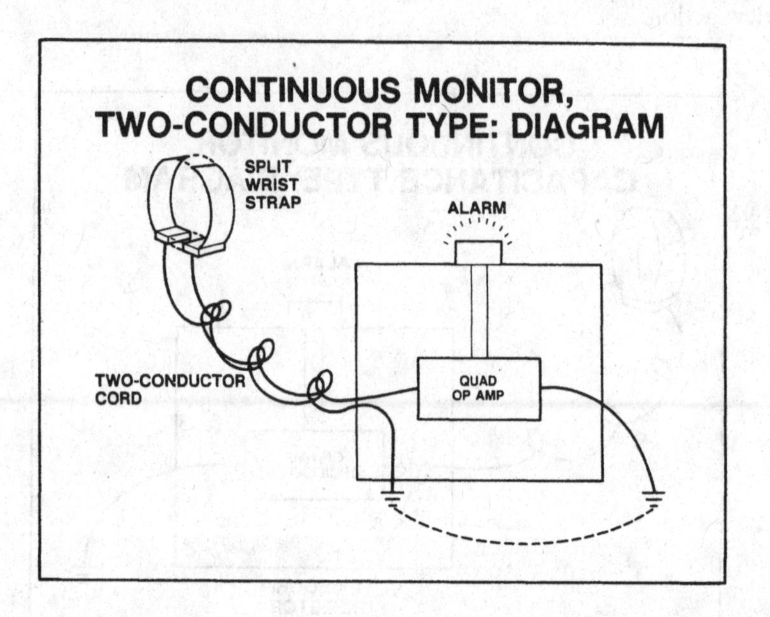

This is a very simplified representation of the Semtronics prototype. The commercial system, Part No. EN435S, will comprise a Sentinel monitor, a dual coil cord, and a wrist strap.

ADVANTAGES OF TWO-CONDUCTOR TYPE CONTINUOUS MONITOR

- FAILURE PROBABLY IN ONE CONDUCTOR, NOT BOTH, SO ESD-SENSITIVE ITEMS NEVER JEOPARDIZED
- CHECKS GROUND CONNECTIONS AS WELL AS OPERATOR
- NOT FOOLED BY HIGH-CAPACITANCE OBJECT TOUCHING STRAP
- WHEN BOTH CONDUCTORS ARE UNPLUGGED FROM BOX, MECHANICAL SWITCH SHUTS OFF ALARM, EG, WHEN OPERATOR TAKES BREAK

At least one maker of the capacitance-type monitor expects to add a feature to check ground connections. Also, the problem of the strap being deceived by a high-capacitance object may possibly be overcome. With these improvements, the capacitance-type monitor would more nearly equal the two-conductor type, so that a choice between the two might largely depend on cost.

Cost effectiveness is not easy to calculate at this point, since the costs of periodic inspection and of hardware damaged due to lack of continuous monitoring must enter the equation. In general, periodic checks may be adequate for low-cost products, but continuous monitoring is indicated for expensive assemblies. After all, most of us consider the wrist strap our first line of defense, and if it fails we have dropped our guard, however briefly, in the absence of continous monitoring.

CONCLUSIONS

- WRIST STRAP FAILURE CAN EASILY CAUSE DAMAGE OF ESD-SENSITIVE DEVICES
- RESISTANCE TEST METHODS ARE DESCRIBED
- 3 FAILURES (RESISTANCE OVER 10 MEGOHMS) IN 36 CHECKS OF STRAPS IN PLANT
- LOOSE BANDS AND DRY SKIN RAISE RESISTANCE
- CONTINUOUS MONITORING PREFERRED
- TWO-CONDUCTOR MONITOR PREFERRED OVER CAPACITANCE TYPE
- TWO-CONDUCTOR MONITOR PROTOTYPES TO BE TESTED IN PLANT
- HIGH-COST ITEMS, EG, $1000 MODULES, DESERVE CONTINUOUS MONITORING

References

1. A. P. Hahl, "A Wrist Strap Life Test Program," *Reliability Analysis Center EOS/ESD Symposium Proceedings,* pp. 94–96, Philadelphia, PA, 1984.
2. J. M. Kolyer, W. E. Anderson, and D. E. Watson, "Hazards of Static Charges and Fields at the Work Station," *Reliability Analysis Center EOS/ESD Symposium Proceedings,* pp. 7–19, Philadelphia, PA, 1984.
3. J. R. Huntsman and D. M. Yenni, Jr., "Test Methods for Static Control Products," *Reliability Analysis Center EOS/ESD Symposium Proceedings,* pp. 94–109, Orlando, FL, 1982.

Paper No. 6

Presented at the 8th Annual Electrical Overstress/Electrostatic Discharge Symposium, Las Vegas, Nevada, September 23–25, 1986, Sponsored by EOS/ESD Association and ITT Research Institute. *EOS/ESD Symposium Proceedings*, EOS-8, 1986, page 111.

METHODOLOGY FOR EVALUATION OF STATIC-LIMITING FLOOR FINISHES

John M. Kolyer and Dale M. Cullop

Rockwell International Corporation
Autonetics Strategic Systems Division
Electronics Operations
3370 Miraloma Avenue
Anaheim, CA 92803

Hanson Loran Chemical Co., Inc.
6700 Caballero Blvd.
Buena Park, CA 90620

Note: The following is a condensed version of this paper. See the *Symposium Proceedings* for the complete text.

Summary

Methodology is described for evaluating floor finishes which limit static buildup on personnel wearing ordinary shoes and drain charges from personnel wearing conductive footwear or heel grounders. Procedures include triboelectric charging of various synthetic shoe sole materials and an accelerated scrubbing test to predict durability. The objective is a low-charging, easily maintained, and cost-effective product.

Introduction

Several commercial ESD-control floor finishes were evaluated in mid-1985. Some were unsatisfactory in limiting static generation, while others had questionable floor finish properties such as scrub resistance; in general, non-ESD properties seemed compromised by antistat addition. All the products were judged too expensive to be cost-effective. These deficiencies prompted our study.

Our objective was a low-charging, easily maintained, and cost-effective acrylic coating, and the search involved development of methodology for evaluating commercial finishes and perfecting new and better ones. The emphasis was on simple, relatively inexpensive equipment and realistic performance tests.

We prefer the term "static-limiting" because this property, not conductivity per se, is what is desired if ordinary shoes are worn. "Static dissipative" and "antistatic" refer to surface resistivity ranges, and surface resistivity correlates imperfectly with triboelectric charging of shoe soles, as will be seen. "Static-preventing" would be a misnomer or false claim because all antistatic materials permit some charging, especially of certain "problem" synthetics.

When conductive footwear or heel grounders are prescribed, low surface resistivity is the key ESD property, and tests for triboelectic charging are superfluous. However, our methodology remains pertinent because it predicts durability in terms of surface resistivity as well as triboelectric charging propensity.

Floor Finishes Tested

The nine commercial products tested represented most of these offered in the U.S. in mid-1985. These were acrylic, modified with wax in some cases, with a minimum solids content of 15–24.5 percent and pH values in the 8–10 range. Note that deficiencies in these products may have been more or less corrected after the time of our testing, because floor finish formulations can be changed at a moment's notice. Also, several new products have appeared. Hence this paper is concerned with general methodology rather than evaluation of specific products.

The four experimental products tested prepresent stages in the evaluation of a new commercial product which was a wax-modified acrylic with a nominal solids content of 20 percent and a pH of 9.7.

Grounded aluminum (alloy 6061, temper T6) was included in the tests as a reference material to demonstrate that a conductive floor surface can charge synthetic shoe soles. Thus a reduced surface resistivity is necessary but not sufficient.

Experimental Methods

The various experimental methods are shown in Fig. 1 through 7. Fig. 8 shows that the actual voltage in the walk test can deviate considerably from the meter reading because of inertia of the needle. To be conservative, we multiply meter readings by a factor of 2.

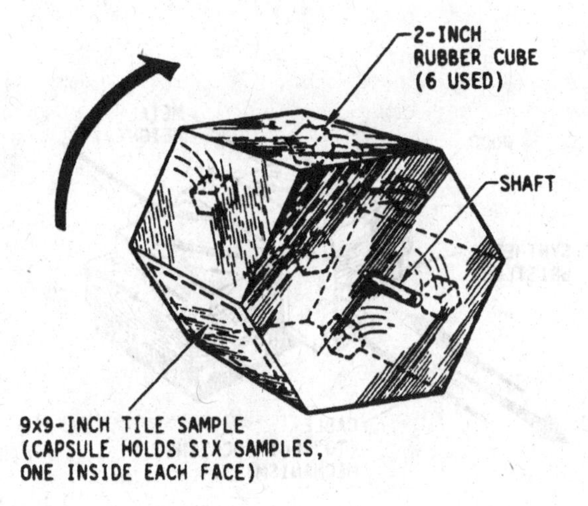

Fig. 1. Heel mark test.

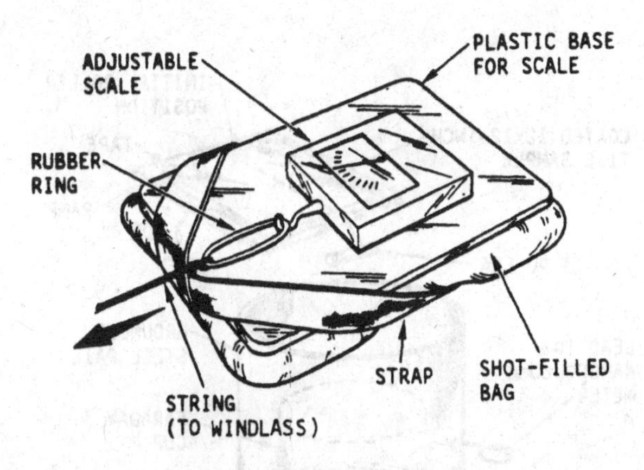

Fig. 2. Topaka slip test.

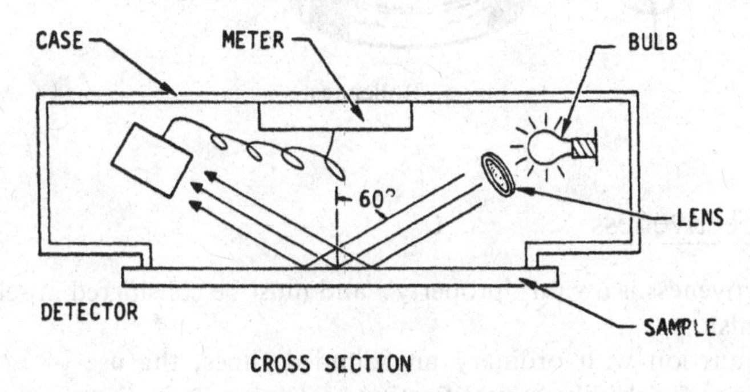

Fig. 3. Gloss test.

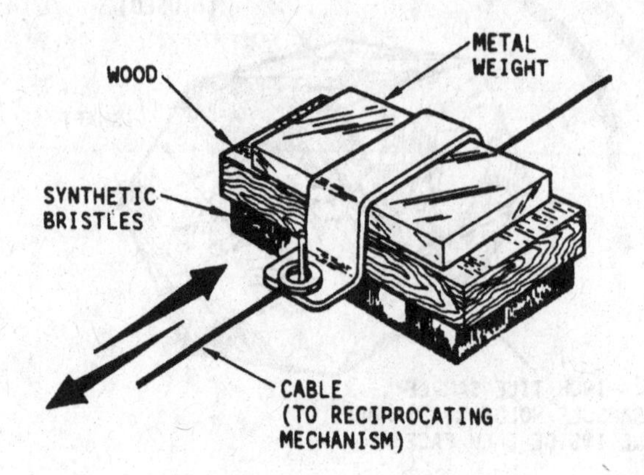

Fig. 4. Detergent or water scrub test.

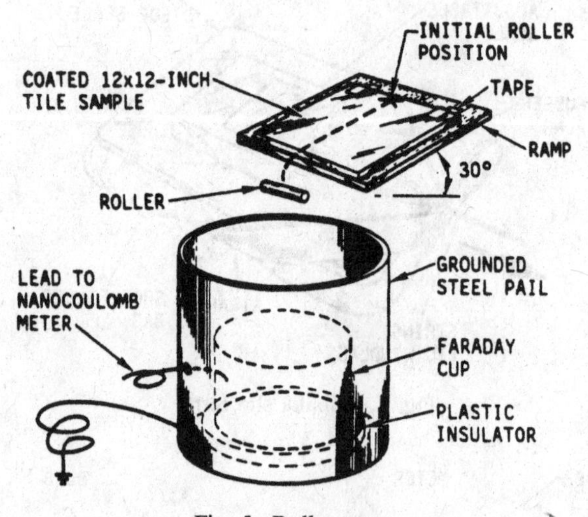

Fig. 5. Roller test.

Cost Effectiveness

Cost effectiveness is a vital "property" and must be considered in selecting a floor finish.

In conjunction with ordinary ungrounded shoes, the use of a static-limiting floor finish, like humidification and room ionization, is a supplementary procedure, as opposed to primary ESD control methods such as

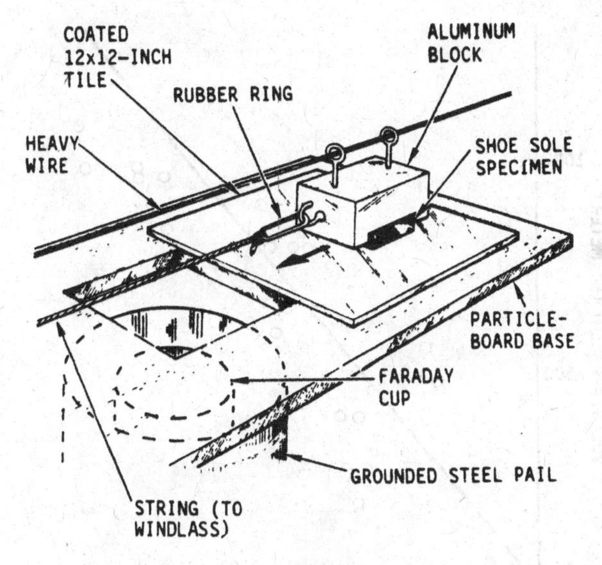

Fig. 6. Drag test.

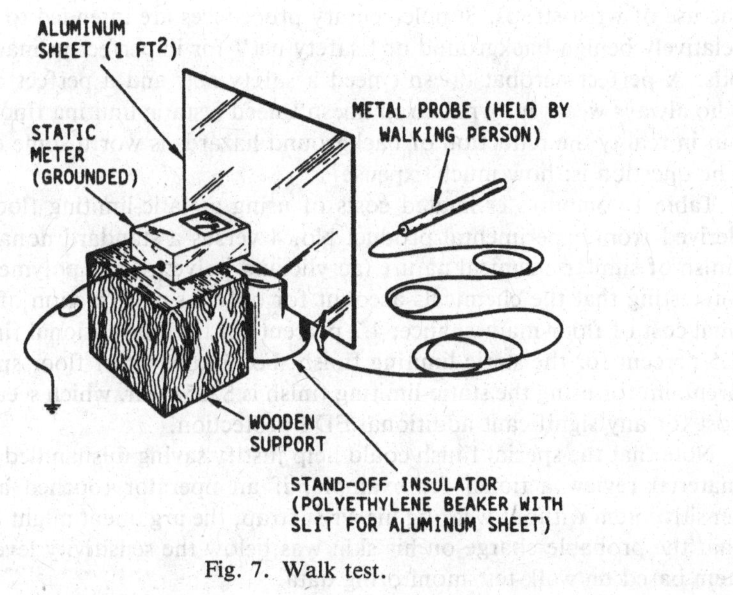

Fig. 7. Walk test.

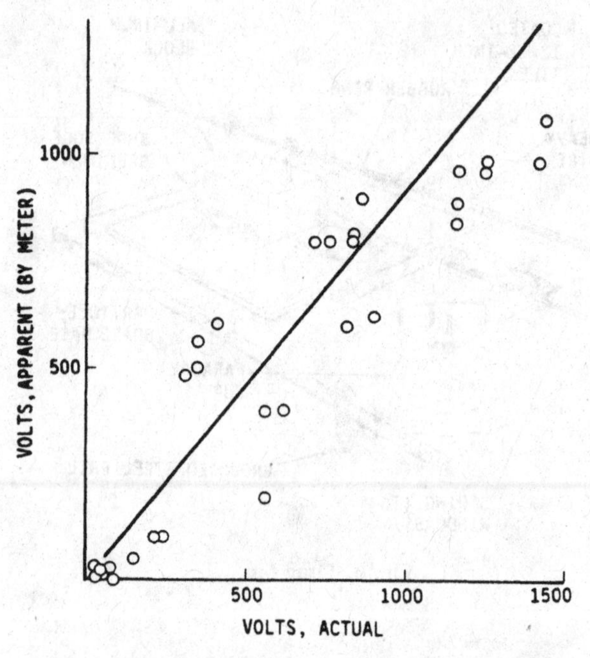

Fig. 8. Apparent versus actual voltage in walk test.

the use of wrist straps. Supplementary procedures are intended to create a relatively benign background or "safety net" for imperfect primary methods. A perfect acrobat doesn't need a safety net, and a perfect operator who always wears his wrist strap doesn't need a static-limiting floor finish, but in reality the reduction of background hazards is worth some expense. The question is: how much expense?

Table 1 compares estimated costs of using a static-limiting floor finish derived from experimental product No. 4 versus a standard nonantistatic finish of similar chemical nature (acrylic and polyethylenic polymers). It is interesting that the chemicals account for only a small fraction of the annual cost of floor maintenance: 4.2 percent for the conventional finish and 6.5 percent for the static-limiting finish. For 10,000 ft^2 of floor space, the premium for using the static-limiting finish is $235/year, which seems a low cost for any significant additional EDS protection.

Note that the special finish could help justify saving mishandled items in material review actions; for example, if an operator touched an ESD-sensitive item without wearing his wrist strap, the argument might be made that the probable charge on his skin was below the sensitivity level of the item based on walk-test monitoring data.

Table 1. Estimated Annual Costs of Using Conventional versus Static-Limiting Acrylic Floor Finish.

	COST, $/10,000 FT^2	
	CONVENTIONAL FINISH, $10/GALLON	STATIC-LIMITING FINISH, $15/GALLON
Finish (8 Coats @ 5 Gallons/Coat)	400	600
Neutral Cleaner (30 Gallons)	165	N/A
Antistatic Neutral Cleaner (30 Gallons)	N/A	300
Maintenance* (5 h/day @ $10/h, 5-Day Week)	13,000	13,000
Totals	13,565	13,900
Premium	0	235

*Scrubbing/recoating, restoring, damp-mopping, dust-mopping, and possibly spray-buffing. These operations are done piecemeal and intermittently, as required.

An advantage of the use of static-limiting floor finish with ordinary shoes as a supplementary procedure is that no disturbing or even visible change is made in the work environment. In contrast, humidification can cause discomfort due to "mugginess," and room ionization has been known to create anxiety about its imagined electrical effects on people. However, thorough testing must assure that voltages are in fact sufficiently controlled with the great variety of shoe soles which personnel may wear. For example, certain vinyl compounds may be "problem" materials, as indicated by high charges in both the walk test and laboratory tests.

Static-limiting floor finishes may be applied to conductive tile without destroying its grounding capability. For example, two coats of the product derived from experimental product No. 4 increased the surface resistivity of a popular brand of conductive tile from 1×10^7 to 3×10^7 ohms/square and increased the resistance to ground (through a 5-lb, 2.5-inch-diameter NFPA 56A electrode) from 7×10^6 to 3×10^7 ohms at 10 volts. These threefold or fourfold increases do not prevent rapid electrical drainage from conductive footwear, and the finish provides wear-resistance and an attractive gloss.

Conclusions

1. Nine products commercially available in mid-1985 were evaluated to illustrate our methodology and found to be generally unsatisfactory in performance and too high in cost. Resistance to scrubbing and foot traffic

seemed compromised by antistat addition. However, formulations are easily changed, and today's versions may be improved.

2. The products tested were not literally "zero-charge." They limited triboelectric charging of shoe soles but did not stop it. As of May 1986, some suppliers felt that ordinary, ungrounded shoes could be worn without charging enough to damage ESD-sensitive devices, but other felt that conductive footwear or heel grounders were needed to drain charges including those created by movement of clothing.

3. The "acid test" of a static-limiting floor finish is the measurement of charges on people who walk on it. When ordinary shoes are worn without conductive straps, surface resistivity correlates imperfectly with charging and can be misleading.

4. The ever-changing shoe sole market presently includes a variety of synthetics such as polyurethane, vinyl, ethylene-vinyl acetate, and styrene-butadiene rubber (SBR). These materials vary in susceptibility to triboelectric charging; vinyls are high and polyurethanes tend to be low (hence they are often used, with additives, in ESD-control footwear). Also, different synthetics may respond in varied and unexpected ways to different finishes. Therefore, an averaging effect is desired, and screening of candidate finishes should include several synthetics, not just Neolite (an SBR compound) as in AATCC Test Method 134-1979.

5. Compared to synthetics, leather soles are a minor problem. They hold only 15 percent of the market, are usually used only on expensive shoes, and in general triboelectrically charge much less than synthetics.

6. In laboratory tests, cylinders of various synthetics were rolled down a ramp of coated tile, or pieces cut from actual shoe soles were dragged across the surface; in both cases the specimens fell into a Faraday cup to measure the charge. The results roughly agreed with voltages observed on people with random shoe soles when a *variety* of synthetics and *several* people were measured to get average effects.

7. Based on these ESD tests, methodology is proposed for developing/evaluating new products in the laboratory. Basic floor finish properties such as slip, gloss, and scrub resistance are included. The test equipment is simple and easy to use.

8. Correlation of laboratory tests with a floor test in a busy supermarket was good for floor finish properties: durable finishes were distinguished from those subject to hazing and scuffing. However, ESD properties deteriorated in all cases due to spray-buffing under the wrong conditions (coatings not thick enough). It is recommended that spray-buffing of static-liming floor finishes be done with caution. A thin restorer coat keeps most of the luster while reviving ESD properties.

9. With proper maintenance, including mopping with a solution of anti-

static cleaner, a good static-limiting floor finish should control the charging of walking personnel to 200 volts maximum (peak voltage of spikes) for most, but not necessarily all, synthetic soles at 50 percent relative humidity and 72°F. An experimental finish was successful at 33 percent relative humidity and 72°F in limited testing.

10. Our tests were done in the vicinity of 50 percent relative humidity, which proved acceptable for comparative purposes (screening of candidate finishes), but final on-the-floor testing must involve the lowest relative humidity to be met in practice. Literature data suggest a threefold rise in charging of rubber soles when the relative humidity drops from 50 to 15 percent at 72°F, but the increase could be much more.

11. Static-limiting floor finishes lower triboelectric charges on rolling equipment as well as walking people. For example, annoying charges on supermarket shopping carts with polyurethane wheels were reduced to an innocuous level.

12. The use of static-limiting floor finish with ordinary, ungrounded shoes is only a supplementary procedure which provides a "safety net" for primary procedures such as the use of wrist or ankle straps. Therefore, the added expense of using the finish must be moderate for this backup procedure to be cost-effective. An example is given in which switching from a conventional to a durable, easily maintained static-limiting finish costs a premium of only $235/year/10,000 ft^2 of floor. On conductive tile, the finish permits grounding while providing wear-resistance and gloss. Since some sort of finish must be used, why not a static-limiting or dissipative one at little added expense?

Note: This paper appeared as an article in *EOS/ESD Technology Magazine,* April 1987, page 20:

Getting ESD Floor-Finish Tests Rolling Toward Standarization

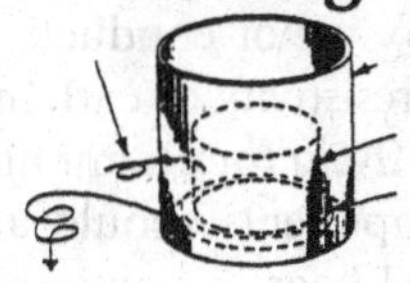

By **John M. Kolyer,** Rockwell International and
Dale M. Cullop, Hanson Loran Chemical Co.

These tests are good candidates for a standard method of evaluating the static-limiting properties of floor finishes.

Paper No. 7

This paper is reprinted, by permission, from *EOS/ESD Technology*, October/ November 1987.

TOTE BOX MATERIAL: HOW GOOD IS IT?

J. M. Kolyer, W. E. Anderson and D. E. Watson

Rockwell International Corp.,
Electronic Operations,
Anaheim, CA

Until recently, all tote boxes for ESD control were made of polyolefin and either topically applied antistat, extruded-in antistat, or extruded-in graphitic (conductive) carbon. Each of these materials has its limitations.

Topically applied antistat is only an expedient because the antistat wears off after some undetermined period of use. Also, a wall of plain polyolefin, even with an antistatic surface, provides little Faraday-cage shielding protection from external static fields or discharges (Table 1).

Extruded-in antistat provides a longer life than topically applied antistat because it continues to bleed to the surface for some time, creating a weakly conductive sweat layer from atmospheric moisture. However, handling, heat, contact with paper products, or exposure to solvents will eventually deplete all the antistat (Ref 1). Again, shielding is poor.

Extruded-in conductive carbon offers the advantage of permanence, but it has several problems. A carbon-loaded polyolefin tote-box wall is conductive enough to endanger people; a current of over 100 mA, which is usually fatal (DOD-HDBK-263), can be carried at 110 V (Table 2). If the conductive box itself is charged it is more dangerous to devices than antistatic or nonconductive boxes (Table 3 and Ref 2).

A high-conductivity surface is also dangerous to devices (Ref 3), especially if the operator should be charged and touches a sensitive lead to the box (Table 4).

However, the wall of a carbon-loaded polyolefin box is not conductive enough to be a good Faraday cage; note that the volume resistivity of carbon-loaded polyolefin is on the order of $10^2 \Omega$-cm versus 10^{-6} Ω-cm for aluminum foil. Refs 3 and 4 agree that highly ESD-sensitive components should always be protected by metal-foil bags, not carbon-loaded bags or bags with thin, see-through metallization. This conclusion is applicable to other containers such as boxes. Furthermore, carbon-loaded polyolefin imparts high triboelectric charges to nonconductors stroked on it (Table 5 and Ref 1) and sloughs conductive particles that could fall into open microelectronic devices and cause shorts.

Table 1. Shielding/Discharge Test.

TOTE BOX TYPE	CAPACITANCE (pF)	RESISTANCE (Ω)	V	MOSFET DAMAGE*
Antistatic or carbon-loaded polyolefin (0.140 in. wall)	Human	Human	−8000	3/3 (0S)
Same as above	97	1500	+5000	2/2 (2S)
Same, but with air gap**	97	1500	+15,000	2/2 (2S)
Carbon-loaded polyolefin as above, but lined with 0.0003-in. aluminum foil	97	1500	+5,000	0/3
Corshield containing one layer of 0.00025-in. aluminum foil	Human Tesla coil	Human Tesla coil	−8,000 35,000	0/10 1/5 (1S)
Corshield folded to give two layers of the foil	Tesla coil	Tesla coil	35,000	0/5
Vinyl (0.011 in.) on 20-gauge aluminum sheet (0.0375 in.)	Tesla coil	Tesla coil	35,000	0/5
Sandwich of aluminum screen sandwiched between 0.060-in. sheets of "Forbon" hard vulcanized fiber (NVF Co., Container Div.)	Tesla coil	Tesla coil	35,000	0/5

*For example, 2/5 (1S) would mean that five MOSFETs were tested, two were damaged and one of those damaged was shorted.
**1-in. gap between each electrode and inner surface of box.

Table 2. Current Carried by Tote Boxes.

TOTE BOX TYPE	RESISTANCE (Ω)		CURRENT (mA)	
	100V	200V	110V	220V
Antistatic	2×10^{9}	2×10^{9}	5×10^{-5}	1×10^{-4}
Carbon-loaded polyolefin	(950 Ω at 1.5 V)		120	230
Corshield	5×10^{8}	4×10^{8}	2×10^{-4}	5×10^{-4}
Vinyl-aluminum sheet	$> 10^{12}$	2×10^{11}	$< 10^{-7}$	1×10^{-6}

Table 3. Damage to MOSFETs by Grounded Operator Touching Charged Tote Box.

TOTE BOX TYPE	VOLTAGE OF CHARGED TOTE BOX			TOTAL MOSFETs DAMAGED
	+500	+1000	−8000	
Antistatic	0/1	0/5	1/5	1/15
Carbon-loaded polyolefin	1/5	2/5	3/5	6/15
Corshield	1/5	0/5	1/5	2/15
Vinyl-aluminum sheet	0/5	0/5	1/5	1/15

Table 4. Damage to MOSFETs by Charged Operator

TOTE BOX TYPE	MOSFET DAMAGE
Antistatic	0/5
Carbon-loaded polyolefin	3/5 (15)
Corshield	0/5
Vinyl-aluminum sheet	0/5
Bare aluminum sheet	4/5 (35)

Table 5. Triboelectric Charging Data.

TOTE BOX TYPE	V ON COUPON			nC ON COUPON		
	FR-4	ACRYLIC	AL	FR-4	ACRYLIC	AL
Antistatic	70	700	0	0.6	6.4	0.0
Carbon-loaded polyolefin	400	800	0	3.8	8.5	0.0
Corshield	200	200	0	1.0	2.9	0.1
Vinyl-aluminum sheet	200	200	800	3.3	2.6	2.5

Looking at New Materials

The basic defect of conventional boxes is that they are insufficient Faraday cages. So, two multilayer designs with permanent ESD properties have been investigated: fiberboard-foil and vinyl-metal sheet. Multiple layers are necessary because a homogeneous wall is not capable of providing both a safe antistatic or nonconductive surface and Faraday cage protection.

Commercially available fiberboard-foil construction ("Corshield" by Conductive Containers Inc.) consists of aluminum foil sandwiched between layers of fiberboard. The fiberboard has a naturally antistatic surface but is covered with an antistatic coating to seal in sulfur-containing impurities that might tarnish silver-plated leads. The foil provides superior Faraday-cage protection. For example, a discharge from a key held by a person charged to 8000 V caused no damage to field effect transistors (MOSFETs), whereas an impractically heavy wall (0.140 in.) of carbon-loaded polyolefin allowed transistor damage (Table 1).

Even the Tesla coil test was passed when the Corshield box was folded to give two layers of foil (Table 1). In previous work (Ref 5), only constructions with heavy foil or metal screen passed this test. Note that the Tesla-coil test is the worst case in electrical stress and positioning of the device in the box. However, this test is not the worst case in statistical significance because only five parts are tested for a "pass" rating. Furthermore, our Tesla-coil test does not use worst-case acceptance criteria because subtle damage may not be seen by the curve tracer, and the MOSFETs used are sensitive to 100 to 200 V whereas some new devices may be affected by only 20 volts. If a cost-effective material can pass this test, we would use that material.

Fiberboard-foil costs less than other materials. Also, it can be stored easily as flat sheets and then folded into boxes when needed. Its only major defect is it limited durability, but heavier fiberboard will probably prove sturdy enough for most applications.

The vinyl-metal sheet design should satisfy the market niche requiring extreme durability. The metal, either steel or aluminum, provides high structural strength and is coated on both sides with tough vinyl, e.g., 0.010 in. thick, by either lamination or powder-coating. The resulting nonconductive surface is safe for people. Also, in a contrived charge-device model test (Table 6), the nonconductive surface was even safer for devices than an antistatic surface.

The relatively heavy metal wall, 0.0375-in. aluminum (20 guage), is a virtually impregnable Faraday cage and suppresses the voltage of a static charge, no matter how much the surface may be stroked, so that the box never has an appreciable E field when the metal is grounded via bare metal

Table 6. Charged-Device Model Test

TOTE BOX TYPE	MOSFET DAMAGE
Antistatic	2/5 (1S)
Carbon-loaded polyolefin	2/5 (0S)
Corshield	1/5 (1S)
Vinyl-aluminum sheet	0/5
Bare aluminum sheet	2/5 (1S)

Table 7. Drain Time Test.

TOTE BOX TYPE	V (NEGATIVE) AFTER 1 SEC	V (NEGATIVE) AFTER 5 SEC
Antistatic	8000	8000
Carbon-loaded polyolefin	5000	900
Corshield	2500	600
Vinyl-aluminum sheet with bolt heads for feet	900	600

feet on the bottom. In our test, the effectiveness of draining off charges onto an antistatic bench-top was better for a vinyl-metal sheet box than for a carbon-loaded polyolefin box (Table 7).

However, good contact with the bench surface, aided by flatness of the bottom of the box, can be critical. The slightly flexible Corshield box benefited from being conformable and lying flat, whereas the rigid antistatic or carbon-loaded polyolefin boxes were slightly "dished" (concave) so that only edges or corners made contact.

The nonconductive vinyl surface's only defect is that it can triboelectrically charge conductors, whereas an antistatic or conductive surface cannot (Table 5). However, charging of nonconductive surfaces, e.g., conformally coated circuit-board modules, seems a more important issue, and nonconductive vinyl was less of an offender than carbon-loaded polyolefin (Table 5).

Test Methods

Below are the various test methods that were used to derive the results shown in each of the tables accompanying this article.

Shielding/Discharge Test (Results in Table 1)

An electrode 1.5 m. sq. was taped against the inside surface of one wall of the tote box and connected to the substrate-case lead of a Motorola 2N4351 MOSFET, and a similar electrode was taped against the inner surface of the bottom of the box and connected to the gate lead. Then a discharge was made to the outer wall of the box over the electrode. This discharge was from a charged person holding a key, from a capacitor connected to a resistor and a steel probe, or from a Tesla coil operated for 30 sec. The box sat on a grounded plate during the test.

Current Carried by Tote Boxes Test (Results in Table 2)

NFPA 56A electrodes were placed 1 in. apart on the box surface, and the resistance was read with a Beckman Model L-10 megohmmeter.

Damage to MOSFETS by Grounded Operator Touching Charged Tote Box Test (Results in Table 3)

A grounded operator held the substrate-case lead of a MOSFET and touched the gate lead to the charged tote box resting on a nonconductive plastic stand-off.

Damage to MOSFETS by Charged Operator Test (Results in Table 4)

An operator charged to +1000 volts held a MOSFET (as in Table 1) by the substrate-case lead and touched the gate lead to the grounded tote box.

Triboelectric Charging Test (Results in Table 5)

1.5-in.-square coupons of uncoated aluminum or FR-4 epoxy circuit-board laminate, uncoated or coated with acrylic conformal coating, were shaken in the tote box for 30 sec and then dropped into a Faraday cup or measured with a static field meter. All charges were positive.

Charged-Device Model Test (Results in Table 6)

The capacitor (1300 picofarads) representing a charged circuit board was FR-4 epoxy laminate, 0.096 × 11 × 15 in., copper-plated on both sides with 1-in. unplated borders. The substrate-case lead of a MOSFET (as in Table 1) was connected to the lower side of the capacitor, which was suspended by nonconductive twine. Then the upper plate was charged to +1000 volts, and the gate lead of the MOSFET was touched to the grounded tote box being tested.

Drain Time Test (Results in Table 7)

The tote box, suspended by nonconductive twine, was charged to — 8000 volts, placed on a melamine-formaldehyde laminate table top ($10^{10}\Omega$/sq.) for either 1 or 5 sec, and lifted again; then the field on the box was measured with a static meter.

Summary

Table 8 summarizes our evaluations. Together, the two new multilayer boxes should satisfy all in-plant handling needs for an ESD-control program. These boxes are able to afford secure Faraday-cage protection for even the most sensitive items when electrically continuous lids are being used.

Cost advantage depends on many factors, most notably the number of units produced and the fabrication method. In general, a vinyl-aluminum sheet box would be competitive with an injection-molded carbon-loaded polyolefin box. A vinyl-steel sheet box, though cheaper than aluminum, would be almost three times heavier in the same gauge. In contrast with the above choices, the fiberboard-foil box is inherently inexpensive (Table 8).

Last of all, fiberboard boxes (Corshield) are commercially available at this time; however, vinyl-metal sheet designs are still in the prototype stage.

Another interesting multilayer design is one which utilizes an aluminum

Table 8.

	TOTE BOX TYPE			
CHARACTERISTIC	ANTISTATIC	CARBON-LOADED POLYOLEFIN	CARDBOARD-ALUMINUM FOIL	VINYL-METAL SHEET
ESD shielding	Poor	Poor	Excellent	Excellent
Drain time	Fair	Good	Very good	Excellent
Triboelectric charging				
Nonconductors	Low	High	Low	Low
Conductors	None	None	Almost none	Some
Danger to devices				
Box charged	Low	Moderate	Low	Low
Operator charged	Low	High	Low	Low
Danger to people	Low	High	Low	Very low
Permanence (ESD)	Poor-fair	Excellent	Excellent	Excellent
Durability (physical)	Good	Good	Fair	Good
Cost	Moderate	High	Low	High

screen or foil sandwiched between layers of hard vulcanized fiber (Table 1). The vulcanized fiber material is naturally antistatic, even at low humidity (5×10^{11} Ω/sq after seven weeks' storage at 72°F and 12% RH). As in the case of the vinyl-metal sheet box, the commercial success of this design would depend upon the development of a practicable fabrication method, but both appear to be excellent alternatives for future ESD control.

References

1. J. M. Kolyer and W. E. Anderson, "Permanence of the Antistatic Property of Commercial Antistatic Bags and Tote Boxes," *Reliability Analysis Center EOS/ESD Symposium Proceedings,* EOS-5, Las Vegas, NV (1983): 87–94.
2. J. M. Kolyer, W. E. Anderson and D. E. Watson, "Hazards of Static Charges and Fields at the Work Station," *Reliability Analysis Center EOS/ESD Symposium Proceedings,* EOS-6, Philadelphia, PA (1984): 7–19.
3. R. D. Enoch and R. N. Shaw, "An Experimental Validation of the Field-Induced ESD Model," *Reliability Analysis Center EOS/ESD Symposium Proceedings,* EOS-8, Las Vegas, NV (1986): 224–231.
4. G. C. Holmes, P. J. Huff and R. L. Johnson, "An Experimental Study of the ESD Screening Effectiveness of Antistatic Bags," *Reliability Analysis Center EOS/ESD Symposium Proceedings,* EOS-6, Philadelphia, PA (1984): 78–84.
5. J. M. Kolyer and W. E. Anderson, "Perforated Foil Bags: Partial Transparency and Excellent ESD Protection," *Reliability Analysis Center EOS/ESD Symposium Proceedings,* EOS-7, Minneapolis, MN (1985): 111–117.

Paper No. 8

Presented at the 9th Annual Electrical Overstress/Electrostatic Discharge Symposium, Orlando, Florida, September 29–October 1, 1987, sponsored by EOS/ESD Association and ITT Research Institute. *EOS/ESD Symposium Proceedings,* EOS-9, 1987, page 41.

ELECTROSTATIC DISCHARGE (ESD) CONTROL IN AN AUTOMATED PROCESS

John M. Kolyer, Ronald Rushworth, and William E. Anderson

Rockwell International Corporation
Autonetics Electronic Systems
Sensors and Aircraft Systems Division
3370 Miraloma Avenue
Anaheim, CA 92803

Note: The following is a condensed version of this paper. See the *Symposium Proceedings* for the complete text.

Summary

In the robot-based, computer-controlled installation of devices such as OP-AMPs and CMOS on printed circuit boards, an exaggeratedly sensitive model assembly served to locate ESD hazards and later monitor the process for their absence. This "coupon" approach applies uniquely to automation with its freedom from unpredictable operator error.

Introduction

The objectives were (1) to reveal ESD hazards in an automated, robot-based manufacturing process for mounting components on printed circuit boards (PCBs), (2) to remove or control these hazards, and (3) to monitor the process for continued ESD safety.

The process and the ESD-sensitive components of concern will be described. Then ESD damage mechanisms, hazards, and remedies will be reviewed, and possible hazards will be related to process steps. Next, the "coupon principle" of using an exaggeratedly sensitive model assembly to locate and evaluate ESD hazards will be explained and illustrated with specific coupon designs. Finally, test data obtained with coupons will be discussed. Hazards were eliminated to give an ESD-safe process which could be checked periodically by means of fresh coupons.

Description of Automated Process

In 1986, Rockwell International at Anaheim, CA, opened a new Automated Manufacturing Cell (AMC) for placing and reflow-soldering surface-mounted components on PCBs to build modules for guidance systems. The AMC (Fig. 1) is a highly automated, computer integrated, robot-based manufacturing system comprising several work stations arranged in a loop. An Automated Guided Vehicle moves around the cell to shuttle parts between stations. Materials are delivered to the AMC from an Automated Material System (AMS).

With reference to Fig. 1, the assembly steps are: (1) removal of parts in tote boxes from an automated carousel, (2) fluxing, tinning, cleaning, lead-forming, and kitting, (3) dispensing of dots of solder paste at the rate of two per second, (4) dispensing of adhesive and placement of flatpack, radial, and axial components at the rate of one per 18 seconds, (5) curing of adhesive and baking out of volatiles from the solder paste, (6) vacuum oven baking and vapor-phase soldering, (7 and 8) cleaning to remove flux and other contamination as well as solder balls, (9) manual installation of oversized or heat-sensitive components and also rework, and (10) inspection using sonic digitizers. Units 7 and 8 are connected by a conveyor belt.

Work instructions and parts information are stored in a central, fault-tolerant computer system. At each station, a laser bar code scanner is used to enter data. Necessary manufacturing data appear on cathode ray tubes (CRTs). This paperless system allows easy product changeovers, and several products can be built simultaneously. The precision afforded by automation reduces component scrap so that subassembly costs are minimized and quality is maximized.

The relative humidity in the AMC room is controlled above a minimum of 30 percent (at 72°F). This reduces triboelectric charging but of course does not prevent it.

Coupon Principle

An automated process, being free of unpredictable operator error, uniquely lends itself to a "coupon" approach in which an exaggeratedly ESD-sensitive model assembly is passed through the process as a coupon to reveal ESD hazards. Various coupon designs used in our tests all contained a 2N4351 MOSFET. The assumption was that if this MOSFET showed no measurable damage, in terms of gate-source threshold voltage and shape of the curve on a curve-tracer, there would be no subtle damage (not detected in quality control tests) to devices on a real module. Undetected damage

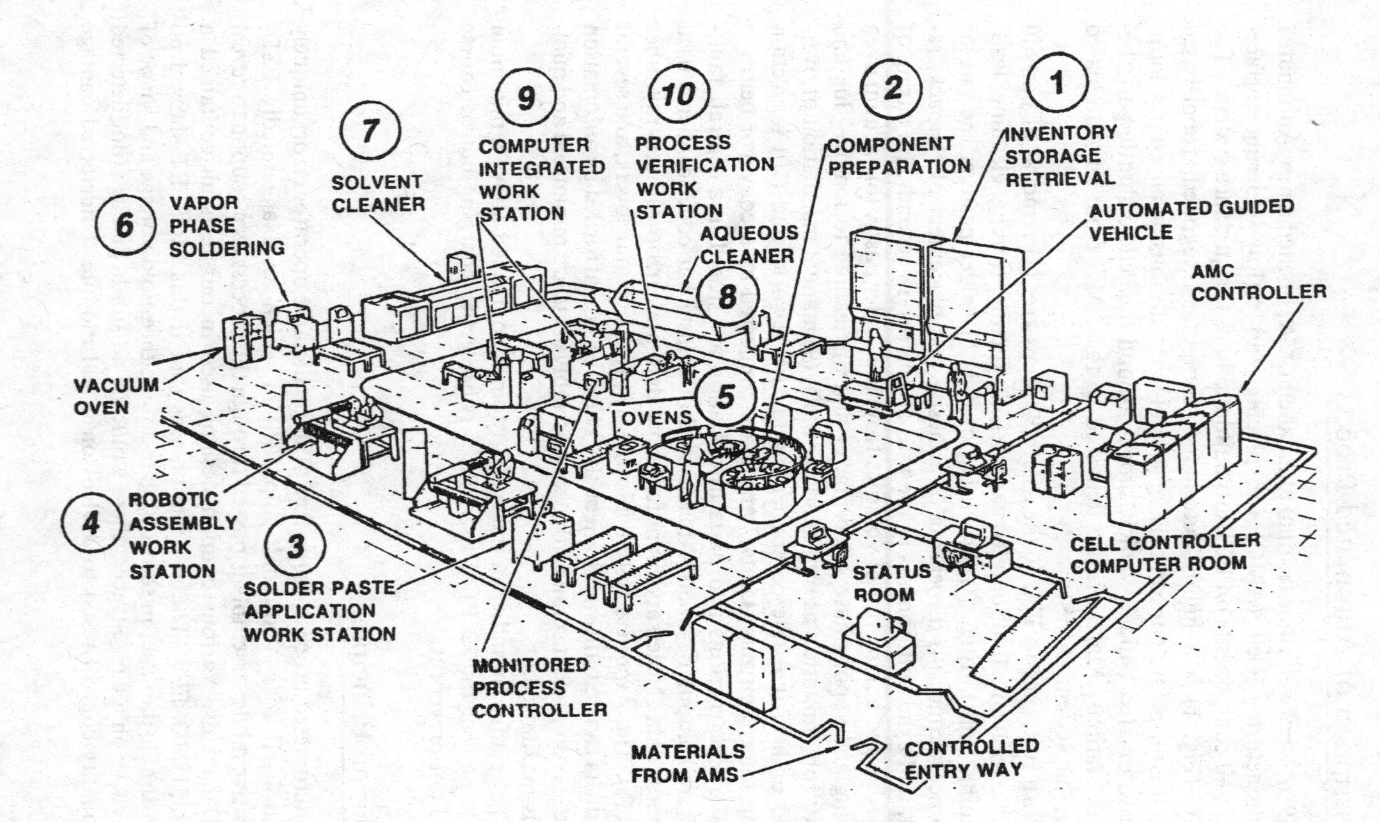

Fig. 1. Automated manufacturing cell.

Fig. 2. Coupon.

Fig. 3. Close-up of encapsulated MOSFET and plugged-in MOSFET.

could lead to latent failure and hence shorter lifetime and reduced reliability. Latent failures are known to occur with CMOS devices.

A coupon is shown in Fig. 2, and a close-up of the MOSFETs on this coupon is shown in Fig. 3.

Experimental Methods

The unedited paper describes how the coupons were tested before and after passing through various process steps. ESD hazards revealed by the coupons were removed, and finally coupons were passed without damage through the entire process and the AMC was declared ESD-safe.

Summary and Conclusions

1. Model assemblies (coupons) which were more ESD-sensitive than real PCB assemblies proved effective in locating ESD hazards in an automated assembly process. This coupon approach was feasible because automation is free from unpredictable operator error.

2. Relevant ESD damage mechanisms were direct injection (DI), the field induced model (FIM), and damage by the static (E) field from the triboelectrically charged board itself (FFB). The most sensitive coupons were damaged by 500 V by DI, 30,000 V at 12 inches by FIM, and 700 V by FFB.

3. The FFB mechanism threatened coupons consisting of bare-board laminates with mounted MOSFETs, but fields on the actual PCBs were minimized by voltage suppression by metallic heat rails and ground planes.

4. Some proposed damage mechanisms never materialized. Triboelectric charging by hot, blowing air was experimentally discredited, and flowing electronic heat-transfer liquid in the vapor-phase soldering process did not damage the most ESD-sensitive coupons.

5. One coupon design was a simulated resistor for handling by robots in the component preparation and adhesive-bonding steps, while other designs were simulated PCB assemblies for passing through the rest of the process. Designs differed in ESD-sensitivity, and when a more sensitive design was damaged, a less sensitive, more realistic design was tried because *all* the designs were far more ESD-sensitive than real PCB assemblies.

6. ESD damage to coupons led to corrective actions. These included eliminating synthetic brushes, removing supplier-installed grounding wires mixed with brush bristles, and making sure all machinery was grounded.

7. The test program culminated in successful passage of five replicate coupons through the entire process. Monitoring would be done in the same way.

8. In conclusion, the empirical coupon approach is recommended for revealing ESD hazards in automated processes and assuring safety after hazards are removed.

9. However, the coupon must be demonstrably more sensitive than actual components to all pertinent ESD damage mechanisms. Introduction of a new component, or a change in the process, could necessitate coupon redesign.

10. Finally, periodic checks with coupons are necessary because continued ESD safety is not guaranteed. For example, a ground connection might fail and machinery parts might be charged by electrical leakage.

Note: This paper was presented as an article in *EOS/ESD Technology Magazine,* August/September 1988, page 10. A good summary was included:

Controlling ESD with Coupons

by John M. Kolyer, Ronald Rushworth and William E. Anderson,
Rockwell International Corp., Autonetics Electronic Systems,
Sensors and Aircraft Systems Div.

A comprehensive and detailed overview of Rockwell International's use of coupons to detect and manage EOS/ESD hazards in an automated assembly facility.

Human unpredictability can cause many static problems, but humans also monitor and control ESD. Here's how Rockwell International uses coupons to detect and manage EOS/ESD hazards in an automated assembly facility where few humans need go.

Editor's Note: Although automated assembly eliminates many of the problems caused by the movement and mistakes of human beings, there remains the problem of detecting and controlling ESD in the absence of human supervision. In a limited way, ESD control becomes an exercise in remote sensing.

In order to spot areas of ESD vulnerability during automated assembly and to assure high quality despite a much reduced human work force, Rockwell International used the methods described below. They are based on the coupon principle, wherein a subassembly more sensitive to ESD than those to be manufactured stands in for the latter during qualification and requalification of the facility.

Paper No. 9

Presented at the 10th Annual Electrical Ovestress/Electrostatic Discharge Symposium, Anaheim, California, September 27-29, 1988, Sponsored by EOS/ESD Association and ITT Research Institute. *EOS/ESD Symposium Proceedings*, EOS-10, 1988, page 99.

CORROSION AND CONTAMINATION BY ANTISTATIC ADDITIVES IN PLASTIC FILMS

John M. Kolyer and Jack D. Guttenplan

Rockwell International Corporation
Autonetics Electronic Systems
Sensors and Aircraft Systems Division
3370 Miraloma Avenue
Anaheim, CA 92803

Abstract

One commercial brand of MIL-B-81705, Type II film contained organic acid and caused corrosion of solder-coated device leads on circuitry. However, solderability was unaffected in accelerated tests. Even acid-free antistats can stresscrack polycarbonate, fog instrument mirrors, weaken adhesive bonds, and discolor epoxy paint. These problems are reduced by a new generation of Type II films.

Introduction

Contamination of surfaces by antistats from antistatic films can cause (1) corrosion of solder (when the antistat contains organic acid), (2) stress-cracking of polycarbonate plastic, (3) fogging of instrument mirrors, (4) loss of strength of adhesive bonds, and (5) discoloration of epoxy-polyamide paint. Other advese effects, e.g., contamination of gyroscope balls, have been discussed in the literature.[1]

Corrosion of Solder

In early 1987 a Government-Industry Data Exchange Program (GIDEP) Alert[2] described corrosion of solder-coated leads by traces of *n*-octanoic acid in one commercial brand of "pink poly" antistatic film (MIL-B-81705, Type II). Therefore, we ran tests to determine the amount of acid present and the nature of corrosion caused by it.

In our tests, extraction of a suspect (acid-contaminated) bag with refluxing ethanol gave acidity corresponding to 514 ppm by weight *n*-octanoic (caprylic) acid, in agreement with 400–650 ppm found by methylene chloride extraction by other investigators.[3] The acidity of a sample of neat (undiluted) antistat from the manufacturer of the suspect film corresponded to 8.9 wt. % *n*-octanoic acid, and other investigators[3] found up to 15 wt. %. From our acidity data, the wt. % of antistat in the film = (514)(100)/89,000 = 0.6, which is roughly the level the manufacturer said had been added to the polyethylene by blending and extrusion.

Our corrosion testing is summarized in Table 1. Degradation of solderability has been suggested[2,3] but was not found in our tests. When painted with neat acid-contaminated antistat, without aging, leads remained wettable by solder, and strength of solder joints was normal. In long-term tests, circuit-board specimens and resistors with freshly tinned copper leads were stored in suspect bags at elevated temperature and humidity or immersed in moist antistat or moist *n*-octanoic acid at 120°F. The conclusion was that corrosion of packaged items does not occur but is only superficial (dull, gray tarnish) because so little acid reaches the solder surface. There was no effect on solderability except in the drastic test with neat *n*-octanoic acid, where enough corrosion products built up to cause dewetting of solder. When the acid content was very low, e.g., 0.05 wt. % as *n*-octanoic acid in the neat antistat, antistats of the diethanolamide type ("proposed antistat" in Table 1) were noncorrosive, and the cosmetic problem of tarnishing disappeared. Most items already stored in acid-contaminated packaging were left as is, but new items were packed in acid-free containers. Panicky repackaging that could deform leads or cause electrostatic discharge (ESD) damage was avoided.

Stresscracking of Polycarbonate

Polycarbonate, which is notably subject to stresscracking, was attacked by all antistats tested, though to different degrees. Stresscrack agents for polycarbonate include commonplace liquids such as ethylene glycol, heptane, and corn oil in the literature[4] and, in our tests, pink liquid hand soap, No. 30 motor oil, and RMA (rosin, midly activated) solder flux. Therefore, it was not surprising that liquid antistats, which have considerable solvent power, were also stresscrack agents. At 2.4% tensile strain on the surface of $\frac{1}{16}$-inch-thick bent strips of polycarbonate, ethoxylated tertiary amine or diethanolamide-type antistats caused cracks to propagate completely through the plastic within 24 hours. At 0.3% strain, the antistats still caused catastrophic cracking, but the time was much longer: 2 to 7 weeks.

When conventional MIL-B-81705, Type II films or new-generation "dry"

Table 1. Corrosion Study.

GIDEP ALERT E9-A-86-02
ISSUED JANUARY 23, 1987

- ONE COMMERCIAL BRAND OF PINK POLY ANTISTATIC FILM (MIL-B-81705, TYPE II) CONTAINS TRACES OF n-OCTANOIC ACID
- n-OCTANOIC ACID REACTS WITH SOLDER, FORMING TIN/LEAD SALTS
- SOLDERABILITY CAN BE REDUCED

TESTING PROGRAM

- **DESIGNED TO ANSWER FOLLOWING QUESTIONS:**
  - IF OPERATIONAL PARTS OR MODULES ARE STORED IN THIS MATERIAL, WILL THERE BE PHYSICAL OR ELECTRICAL DAMAGE, I.E., CORROSION, IMPAIRED SOLDERABILITY, CONDUCTIVE PATHS?
  - IF THERE IS DAMAGE, WHAT DISPOSITION SHOULD BE MADE OF PARTS AND MODULES PRESENTLY STORED OR PREVIOUSLY EXPOSED?
- **TEST SPECIMENS**
  - PCB SPECIMENS, SOLDER-COATED CIRCUITRY BOTH SIDES
  - PCB MATERIAL COUPONS, COPPER-CLAD BOTH SIDES, SOLDER DIPPED OR SOLDER PLATED AND REFLOWED
  - RESISTORS WITH COPPER LEADS, FRESHLY TINNED
- **ACCELERATED EXPOSURE TESTS**
  - PARTIAL IMMERSION AT 120°F FOR 7 DAYS
    - MOIST n-OCTANOIC ACID
    - MOIST NEAT ANTISTAT (FROM SUSPECT BRAND)
    - MOIST PROPOSED NEAT ANTISTAT
  - HUMIDITY TEST (MIL-STD-202, METHOD 106)
    - PAINT ON THIN FILM
      - n-OCTANOIC ACID
      - NEAT ANTISTAT (FROM SUSPECT BRAND)
      - PROPOSED NEAT ANTISTAT
    - CYCLE RM. TEMP. TO 150°F AT 90-98% R.H., TWICE DAILY FOR 10 DAYS
- **RESULTS (CONTINUED)**
  - LONG-TERM STORAGE TESTS
    - SUSPECT BRAND OF BAGS CONTAINS LOW LEVEL, TYPICALLY 500 PPM, n-OCTANOIC ACID
    - OTHER BRANDS ESSENTIALLY FREE OF ORGANIC ACID
    - CORROSION EFFECTS IN SUSPECT BAGS LIMITED TO DULLING/GRAYING; NO EFFECT ON SOLDERABILITY
    - NO CORROSION/NO EFFECT ON SOLDERABILITY IN OTHER COMMERCIAL BRANDS

- **LONG-TERM EXPOSURE TESTS**
  - 80% R.H AT 100°F
    - SEALED AND UNSEALED BAGS
    - 3-MONTHS EXPOSURE
  - AMBIENT HUMIDITY AND TEMPERATURE
    - SEALED AND UNSEALED BAGS
    - 1-YEAR EXPOSURE
- **EVALUATION METHODS**
  - PCB SPECIMENS
    - ADD SOLDER PASTE AND VAPOR PHASE SOLDER
    - VISUALLY DETERMINE DEGREE OF SOLDER WETTING
  - SOLDER-COATED COUPONS
    - SOLDER DIP
    - VISUALLY DETERMINE DEGREE OF SOLDER WETTING
  - RESISTOR LEADS
    - MEASURE SOLDERABILITY ON MENISCOGRAPH
- **RESULTS**
  - ACCELERATED TESTS
    - NEAT ANTISTAT (SUSPECT BRAND) CAUSED DULLING/GRAYING, NO EFFECT ON SOLDERABILITY
    - PURE n-OCTANOIC ACID FORMED HEAVY CORROSION PRODUCTS, SOME EFFECT ON SOLDERABILITY
    - FTIR ANALYSIS - CARBOXYLIC DERIVATIVES (ORGANIC ACID SALTS) OF TIN AND LEAD
    - PROPOSED NEAT ANTISTAT (ESSENTIALLY FREE OF ORGANIC ACID) HAD NO EFFECT ON SOLDER OR SOLDERABILITY

APPEARANCE OF PCB TEST SPECIMENS AFTER ACCELERATED IMMERSION TESTS

- IMMERSION IN PURE N-OCTANOIC ACID SATURATED WITH WATER - HEAVY CORROSION PRODUCT ACCUMULATION IN LIQUID
- IMMERSION IN SUSPECT BRAND NEAT ANTISTAT - SLIGHT GRAYING IN LIQUID
- IMMERSION IN PROPOSED NEAT ANTISTAT - NO EFFECT

ACTION TAKEN

- PURCHASE OF SUSPECT BRAND SUSPENDED
- PARTS AND MODULES PRESENTLY STORED: LEFT "AS IS" (SOLDERABILITY NOT AFFECTED; REPACKAGING COULD CAUSE ESD DAMAGE, LEAD DAMAGE, CONTAMINATION, ETC.)
- PARTS AND MODULES PREVIOUSLY EXPOSED; LEFT "AS IS" (NO PROBLEM OR FAILURE MODE HAS BEEN ATTRIBUTED TO THIS EFFECT)

films (discussed below) were rubbed on the surface of bent polycarbonate strips at 2.4% strain, the new films caused significantly less cracking than did the "old." The explanation is that the new films have minimal liquid antistat on their surface. Thus they may be safe when the stress level on the polycarbonate surface is very low, but the cautious recommendation is to avoid contacting polycarbonate items with any plastic that contains a liquid antistat.

Fogging of Instrument Mirrors

Fogging of instrument mirrors by outgassed antistat during storage was observed and was simulated by suspending an aluminum plate inside an antistatic bag in an oven at 160°F. After one week, the plate was rinsed with acetone, and the solution was analyzed by infrared spectroscopy. It was found that about 20% of the original antistat on the inner surface of the bag (as determined by rinsing a fresh bag) had migrated as vapor to contaminate the plate. The same level of transfer is estimated to occur in 6 months at room temperature, based on a Cox chart which relates vapor pressure of organic compounds to temperature.

The solution to the fogging problem obviously is to use ESD-protective packaging material free of fugitive, surface-seeking additives. Another example of the undesirability of transferable antistats is the packaging of items such as resistors whose surface must maintain high electrical resistivity. In this instance the weakly conductive "sweat layer" that is desired on the package surface becomes a liability on the surface of the stored item.

Loss of Strength of Adhesive Bonds

Intuition suggests that contamination of surfaces by "greasy" antistat would greatly reduce the strength of adhesive bonds. An experiment was conducted using aluminum (alloy 2024 T3) treated with Forest Products Laboratories etch solution (30 parts by weight deionized water, 10 parts concentrated sulfuric acid, and 1 part sodium dichromate) at 150°F for 20 minutes followed by rinsing with deionized water and drying at 150°F for 20 minutes. Lap-shear specimens were prepared using a 2-mil (0.002-inch) bond line of Scotchweld EC2216 (3M Co.), which is a filled epoxy-polyamide adhesive. These bonded aluminum specimens were cured at 155°F for 15 hours and tensile-tested by a standard method.[5] Before bonding, the surfaces had been (1) left clean, (2) contaminated with a slight amount of antistat by rubbing with an antistatic bag, or (3) liberally

Table 2. Adhesive Bond Strength Data

CONTAMINANT	AV. BOND STRENGTH, PSI	STANDARD DEV., PSI	RELATIVE BOND STRENGTH
None	2588	199	1.00
Antistat from rubbing with suspect bag	2468	357	0.95
Antistat from rubbing with new-generation "dry" bag	2583	100	1.00
Ethoxylated tertiary amine antistat	1954	438	0.76*
Suspect amide antistat containing 8.9% n-octanoic acid	2196	132	0.85**

*About 10% adhesive failure (bare aluminum exposed).
**Trace, perhaps 0.1%, adhesive failure.

smeared with neat antistat. Bond strength data (averages of four determinations) are given in Table 2.

Failures were cohesive (within the adhesive) except as noted in the table. As shown by the relative bond strengths, slight contamination by rubbing with the bags had a negligible effect, and gross contamination by ethoxylated tertiary amine or diethanolamide antistat has surprisingly little effect on bonding with one frequently used structural adhesive. However, other adhesives might behave differently and would have to be tested. Our result merely shows that fugitive antistats do not necessarily cause severe bonding problems.

Discoloration of Epoxy-Polyamide Paint

Instrument panels painted with gray epoxy-polyamide paint (MIL-C-22750, Type I) were discolored by prolonged contact with the acid-contaminated brand of MIL-B-81705, Type II film. The film had been pressed tightly against the surface, and wrinkles in the film corresponded to a pattern of whitish streaks or marbling. Abrasive rubbing compounds failed to remove these blemishes, and waxes did not fully restore gloss. Repainting was necessary.

When drops of liquid were left on a painted surface for 24 hours at 120°F, *n*-octanoic acid caused the most attack (swelling and softening of the paint), followed by the acid-contaminated antistat, a low-acid diethanolamide antistat, and an ethoxylated tertiary amine antistat. The explanation for this order of attack is that the acid was the best paint solvent, the amide was intermediate, and the amine was the poorest solvent.

At room temperature, the acid-contaminated film caused staining in 1 day when a pressure of 5 psi was maintained but no staining in the absence of pressure. A sample of new-generation pink antistatic bubble-wrap, which had minimal antistat on the surface, caused no staining even after 6 months at 5 psi.

Attack on paint probably is rarely noticed because the film is not usually tightly pressed against the bagged item as chanced to happen in our case. Our temporary solution was to use MIL-B-81705, Type I, which of course is antistat-free.

Summary and Conclusions

1. One commercial brand of MIL-B-81705, Type II film was contaminated with *n*-octanoic acid and caused corrosion of solder, as revealed in a GIDEP Alert. However, our corrosion tests indicated that the problem was only cosmetic because solderability was unaffected; in fact, the acid resembles flux, which is also an organic acid. Therefore, panicky repackaging was avoided in our operations. The issue is now past, because the offending product has been removed from the market.

2. Antistat-containing films are suitable for most purposes and have a long history of success in ESD control, but they must be used with discretion because the fugitive, surface-seeking additive can stresscrack polycarbonate, fog instrument mirrors, and discolor epoxy paint. Adhesive bond strength also may be reduced, but the effect was less than expected in tests with a commonly used epoxy adhesive.

3. Liquid antistats have considerable solvent power and can even attack rubber and wire insulation, but in practice the damaging effects are limited by the small amount of antistat transferred by rubbing or volatilization. For example, *n*-octanoic acid in sufficient quantity will destroy solder completely, but a minimal amount of the acid contained in rubbed-off antistat merely dulled solder surfaces.

4. Antistat contamination problems are reduced, but not eliminated, by a new generation of "dry" MIL-B-81705, Type II films with a minimal amount of "acid-free, amine-free" amide antistat on the surface. By "acid-free" is meant negligible acidity, and "amine-free" refers to a low amine content, e.g., 4–6 wt. % diethanolamine. Amines are not corrosive to solder under conditions of use, and ethoxylated tertiary amine has long been used as an antistat.

5. Assuming that their permanence[6] proves adequate, the new-generation bags and bubble-wrap will be a great improvement over the old "wet" or

"greasy" products but are not a panacea because the difference is quantitative, not qualitative. Packaging materials free of liquid antistat still must be used in critical applications such as shipment or storage of polycarbonate items. Potential users should run tests to find the most cost-effective packaging system.

Acknowledgment

The assistance of R. Rushworth, D. R. Violette, A. E. Carmellini, and T. J. Hester in performing chemical analyses and corrosion tests is greatly appreciated.

References

1. M. K. Bernett, H. Ravner, and D. C. Weber, "Electroactive Polymers as Alternate ESD Protective Materials," *EOD/ESD Symposium Proceedings,* pages 115–119, Orlando, FL, 1982.
2. GIDEP Alert No. E9-A-86-02, on "Materials, Plastic, Antistatic," issued January 23, 1987.
3. J. Anderson, R. Denton, and M. Smith, "Antistatic Polyethylene Package Corrosion," *EOS/ESD Symposium Proceedings,* pages 36–40, Orlando, FL, 1987.
4. *Modern Plastics Encyclopedia,* 1978–79, pages 528–529.
5. "Standard Test Method for Strength Properties of Adhesives in Shear by Tension Loading (Metal-to-Metal)," ASTM D1002-72 (reapproved 1983).
6. J. M. Kolyer and W. E. Anderson, "Permanence of the Antistatic Property of Commercial Antistatic Bags and Tote Boxes," *EOS/ESD Symposium Proceedings,* pages 87–94, Las Vegas, NV, 1983.

Paper No. 10

Presented at the 11th Annual Electrical Overstress/Electrostatic Discharge Symposium, New Orleans, Louisiana, September 26–28, 1989, Sponsored by EOS/ESD Association and ITT Research Institute.
EOS/ESD Symposium Proceedings, EOS-11, 1989, page 23.

CONTROLLING VOLTAGE ON PERSONNEL

John M. Kolyer, Donald E. Watson, and William E. Anderson

Rockwell International Corporation
Autonetics ICBM Systems Division
Electronics Operations
3370 Miraloma Avenue
Anaheim, California 92803

and Dale M. Cullop

Hanson Loran Chemical Co., Inc.
6700 Caballero Blvd.
Buena Park, California 90260

Note: The following is a condensed version of this paper. See the *Symposium Proceedings* for the complete text.

Abstract

Controlling voltage on personnel is a vital and demanding requirement in ESD protection. New data show that the allowable resistance to ground must be lower than was thought. Furthermore, wrist-strap systems often fail, so monitoring must be continuous; resistive-type monitors are best. A cost-effective program is described.

Introduction

Controlling voltage on personnel has long been recognized as of primary importance in preventing electrostatic discharge (ESD) damage to electronics. The reason for concern is that operators' fingers often touch leads or contacts of ESD-sensitive (ESDS) devices, and a flow of current from person to device can destroy or "wound" it by the direct injection (DI) mechanism. A "wound," passing undetected in routine tests, may result in latent failure. Thus, improperly grounded people are the principal ESD hazard or at least prime sources of ESD for destroying parts.

A resistance of about 1 megohm is generally agreed upon for current-limiting resistors to protect operators wearing wrist straps. However, questions remain: (1) What should be the maximum allowable resistance to ground (ARTG) for protection of ESDS items as opposed to protection of operators? (2) How can the triboelectric charging of operators be minimized to make the resistance to ground less critical? (3) Should this resistance be checked at intervals or continuously? (4) If a continuous wrist-strap monitor is used, which design is best? (5) How can skin-voltage control and monitoring be made part of a cost-effective program? We have attempted to answer these questions based on experience as well as experimental data.

Experimental Methods

An operator sat at a workbench and charged himself either by stroking a sheet of Aclar film on the bench surface or by shuffling his feet. The resulting data are shown in Fig. 1. Fig. 1 represents worst-case charging conditions, but surveys of operator grounding in manufacturing areas showed many bad cords and instances of high resistance to ground because of operators' dry skin. Therefore, we recommend continuous monitoring with units like the one illustrated in Fig. 2 and 3.

A Cost-Effective Program Including Skin-Voltage Control

We emphasize basic, proven methods of ESD control and deemphasize methods which are of questionable value and may even be part of the problem. The major elements in our suggested program for ESDS items in Class 1 of MIL-STD-1686A are:

1. Operator Disciplines

A sufficiently skilled operator could work with almost no special equipment, and, conversely, the most expensive appliances won't prevent damage by an ignorant or careless operator. Remember that a grounded operator is only the lesser of two evils versus an ungrounded operator because a charged device could be damaged by rapid discharge to the grounded operator's finger (the Charged Device Model or CDM). To maintain their skills, operators must be retrained periodically.

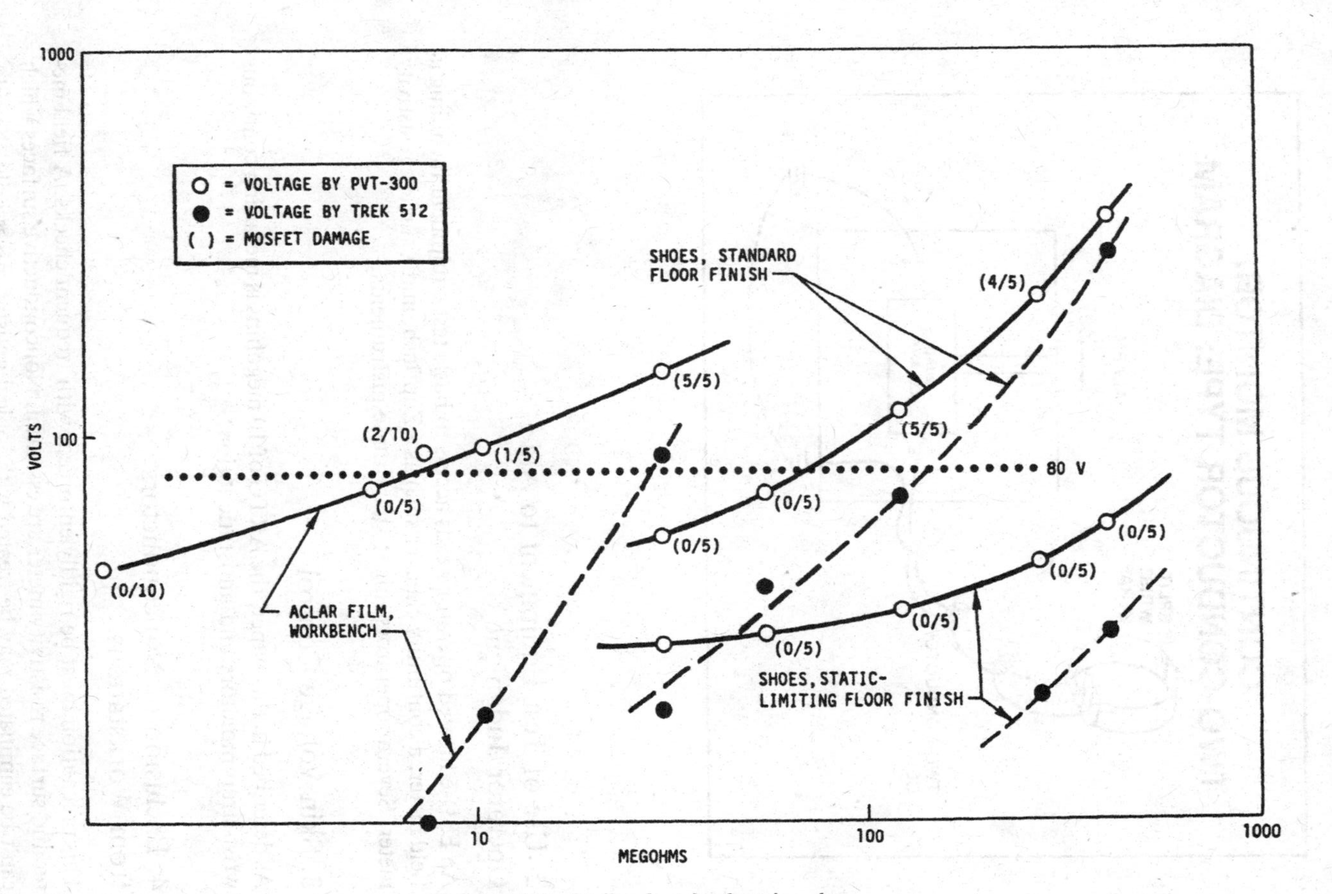

Fig. 1. Plot of triboelectric charging data.

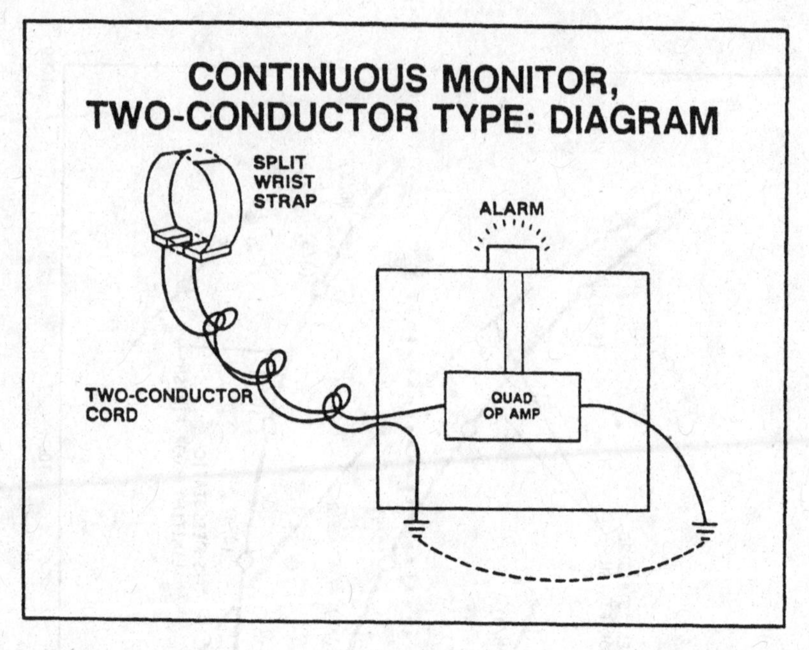

Fig. 2.

2. Use of Test Equipment to Aid Operator Judgment

An ESD-educated operator needs access to basic test equipment including a field meter, a continuity checker such as a Zapflash, and a surface resistivity meter. Several operators might share these instruments.

3. Skin Voltage Control

As described in this paper, the ARTG of 10 megohms is met using continous wrist-strap monitors and antistatic lotion.

4. Exclusion of Nonconductors from Workstations

This precaution must be rigidly enforced, with frequent checks. A field meter and surface resistivity meter are essential. Nonconductive surfaces which can't be eliminated may be treated with topical antistat, necessarily chloride-

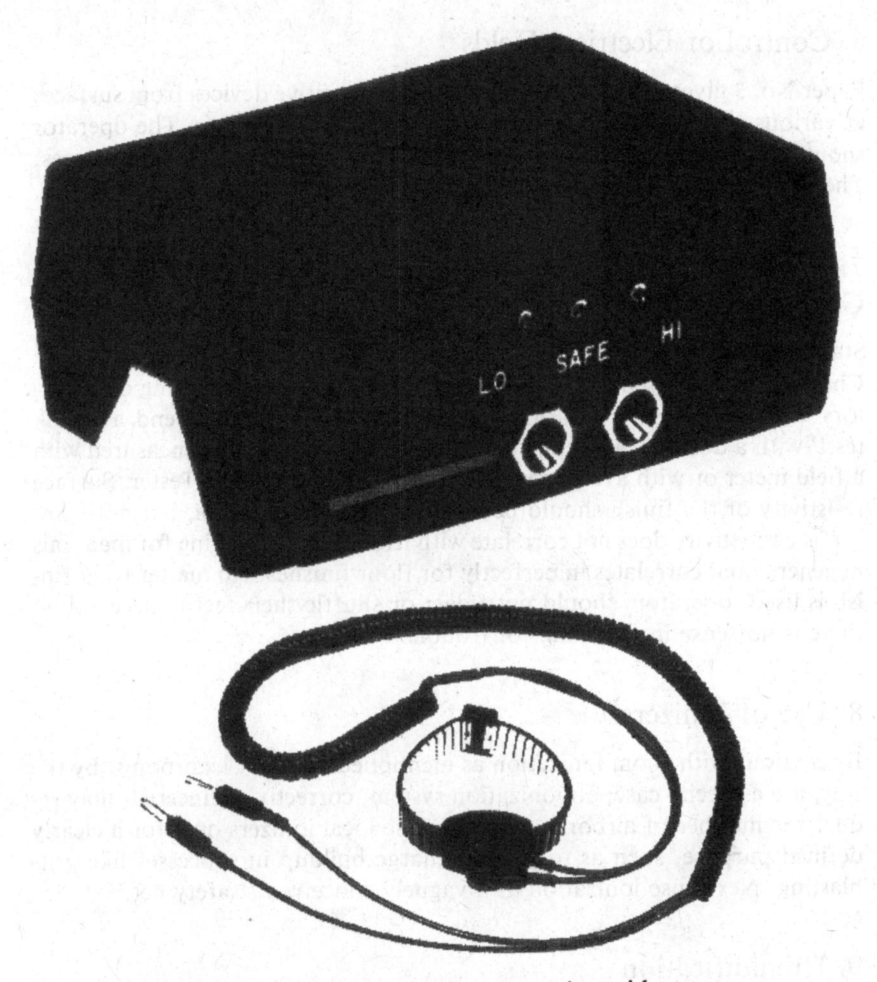

Fig. 3. Two-conductor continuous monitor with strap.

free in some applications, and so labeled. Bench tops should have static-dissipative surfaces and a buried ground plane to suppress voltage.

5. Grounding of All Conductors

This is easily checked with a Zapflash. If a conductor, such as a microscope eyepiece housing, cannot be conveniently grounded, the operator must satisfy himself that it never becomes dangerously charged. The best policy is to ground even small-area conductors.

6. Control of Electrical Fields

Paper No. 3 gives safe distances for 100-volt-sensitive devices from surfaces at various apparent voltages as determined by a field meter. The operator should run frequent checks on his static-safe workstation to locate charges. Then, if these can't be eliminated, he will work a safe distance from them.

7. Minimization of Triboelectric Charging of Operators

Static-limiting floor finish is very cost-effective as seen in Paper No. 6. Choose a brand based on its ability to limit tribroelectric charging of operators wearing shoes with various types of soles. We recommend a "walk test" with a dozen of your personnel: the voltage on them is measured with a field meter or with a Voyager PVT-300 Personnel Voltage Tester. Surface resistivity of the finish should be at most 10^{11} ohms/square, but note that surface resistivity does not correlate with triboelectric charging for materials in general and correlates imperfectly for floor finishes. No matter what finish is used, operators should not fidget or shuffle their feet unnecessarily; there is no sense in "looking for trouble."

8. Use of Ionizers

Be cautious with room ionization as mentioned above. Cleanrooms, by the way, are a special case; an ionization system, correctly engineered, may reduce the number of airborne particles. Use local ionizers only for a clearly defined purpose, such as to control charge buildup in processes like grit-blasting. Never use ionization as a vaguely conceived "safety net."

9. Humidification

In very dry areas, humidification is desirable because it makes antistatic materials with "sweat layers" function better and it reduces triboelectric charging. But don't let it build false confidence, and beware of corrosion or other problems as mentioned above. As with ionization, the use of humidity control must be judicious.

10. Static-Safe Packaging

We recommend "Faraday-cage" packaging with aluminum foil as shielding, because independent studies agree that foil is the only adequate protection from external fields and discharges for highly ESD-sensitive items.

When away from a static-safe workstation, an ESDS item must be in a Faraday-cage container such as a tote box for in-plant handling or a sturdy, suitably cushioned Faraday-cage package for shipment.

11. Elimination of Unnecessary Equipment

Our recommended program is cost-effective because it emphasizes operator skills coupled with basic, clearly useful equipment. Money is saved by eliminating marginally valuable appurtenances such as special smocks, footwear, chairs, or floors. "Buy the basics, forget the frills." Besides being a waste of capital, some frills may be counterproductive by instilling false confidence so that operator disciplines are relaxed.

12. Program Administration

Of course, good record-keeping and day-to-day management are essential. Operators should be encouraged to keep records, e.g., of hazards such as E fields that they've encountered, and to make suggestions. Incidents of hardware mishandling should be discussed with the group. Operators will monitor their individual workstations by using test equipment as recommended above. With alert, knowledgeable operators and an emphasis on the basics, ESD control becomes a simple matter except for silicon wafer processing, paint-spraying, or other processes which involve triboelectric charging beyond the operator's control. Static-limitation must be designed into such processes, e.g., by using conductive solvents or blowing ionized air over surfaces. Automated processing is, of course, out of the operator's hands by definition and requires special approaches such as the use of ESDS coupons to locate hazards. In large companies, a Program Coordinator should be appointed to lead a working group, conduct monthly meetings, brief upper management, etc.

Summary and Conclusions

1. An operator, grounded through various resistances, triboelectrically charged himself by shuffling his feet or by stroking a plastic film (representing a garment sleeve) on a workbench surface. Devices sensitive to 80–100 V were damaged by being touched by the operator at a little over 70 megohms resistance to ground in the foot-shuffling tests and a little over 7 megohms in the plastic film tests. An explanation for the difference involves rapidity of voltage rise, which is greater for the thinner dielectric (film versus shoe soles).

2. Considering such factors as the methodology of the tests, the effect of

relative humidity, and the fact that women's shoe soles tend to be thinner than men's (as used in the test), we recommend an allowable resistance to ground (ARTG) of 10 megohms. This is the maximum, and a suitable minimum for most work is 1 megohm, so as to limit current to well under 1.0 mA. When high voltages are being handled, ground fault circuit interruptors will guard against the remote possibility of an electrically overstressed resistor being carbonized so that it falls below a safe resistance.

3. To minimize triboelectric charging, operators shouldn't fidget or shuffle their feet unnecessarily. Also, a static-limiting floor finish is effective; for example, the voltage on an ungrounded, foot-shuffling operator was reduced from 3300 V with an ordinary finish to 180 V with the special finish. With the special finish, the ARTG could have been increased by a factor of about 8, because the charging of the operator was so low even when he was ungrounded. However, the ARTG must be set assuming normal floor surfaces.

4. The static-limiting floor finish remained effective, though less so, after severe scuffing and scraping to simulate wear. We advise selecting a durable brand of finish which is easily maintained by standard procedures.

5. One-conductor wrist straps show poor reliability. In fact, checks on wrist straps in electronic assembly areas disclosed not only a distrubing rate of cord failure but many instances of excessive resistance to ground caused by loose wrist bands or operators' dry skin. Clearly, monitoring is essential.

6. This monitoring should be continuous because ESD damage can occur in nanoseconds and an ESDS item can be "zapped" during the interval, however brief, between periodic checks. All hardware handled since the last successful check of a failed strap could be subject to a material review action and might even be scrapped! Furthermore, cords sometimes fail in an intermittent way which could go undetected in periodic checks but be revealed in continuous monitoring. We highly recommend these monitors, but some users have been troubled by features such as the audible alarm. Potential buyers should run trials in their own assembly areas.

7. Our preferred monitor design is the two-conductor resistive type which checks ground connections as well as the operator. Also, the two conductors provide redundancy (if one fails, the other keeps working), and the monitor is not fooled by a high-capacitance object touching the strap. The property of interest—resistance—is measured directly. Antistatic lotion is often required for dry skin, and the monitor gives notice when the lotion must be renewed.

8. This paper sketches a cost-effective program, including skin-voltage control, for protecting items in sensitivity class 1 of MIL-STD-1686A. Elements of the program include operator disciplines, test equipment for oper-

ators' use, continuous monitoring of skin voltage, exclusion of nonconductors, grounding of conductors, control of E fields, minimization of triboelectric charging of operators, judicious use of ionization and humidification, static-safe packaging, cost-saving by eliminating unnecessary equipment ("Buy the basics, forget the frills"), and astute program administration.

Index

9 780442 003470